INFORMATION TECHNOLOGY AND MANAGEMENT STRATEGY

INFORMATION TECHNOLOGY AND MANAGEMENT STRATEGY

Editors
Kenneth C. Laudon
Jon A. Turner
Graduate School of Business Administration
New York University

Prentice Hall
Englewood Cliffs, New Jersey 07632

Library of Congress Cataloging-in-Publication Data
Information technology and management strategy.

Bibliography: p.
Includes index.
1. Information technology. 2. Strategic planning.
I. Laudon, Kenneth C. II. Turner, Jon A.
T58.64.I55 1989 658.4'038 88-30726
ISBN 0-13-465022-0

Editorial/production supervision and
interior design: Jacqueline A. Jeglinski
Cover Design: Ben Santora
Manufacturing buyer: Margaret Rizzi

A Division of Simon & Schuster
Englewood Cliffs, New Jersey 07632

Printed in the United States of America
10 9 8 7 6 5 4 3 2 1

ISBN 0-13-465022-0

Prentice-Hall International (UK) Limited, *London*
Prentice-Hall of Australia Pty. Limited, *Sydney*
Prentice-Hall Canada, *Toronto*
Prentice-Hall Hispanoamericana, S.A., *Mexico*
Prentice Hall of India Private Limited, *New Delhi*
Prentice Hall of Japan, Inc., *Tokyo*
Simon & Schuster Asia Pte. Ltd., *Singapore*
Editora Prentice-Hall do Brasil, Ltda., *Rio de Janeiro*

CONTENTS

PREFACE

The articles in this volume were originally given as papers at the New York University Symposium on Information Technology and Strategic Management in June, 1986. The purpose of the Symposium was to draw together researchers who had a keen understanding of the literature and practice of strategic planning, organizational performance, and information technology.

From the beginning we sought to avoid what by 1986 had become a chorus of consultants, vendors, visionaries, and not a few academics singing the praises of information technology. We wanted instead a realistic assessment of both the successes and the failures in the strategic uses of information technology.

Our view in 1986, a view grown stronger with the passage of time, was that the ability of organizations to use information technology for sustained, competitive advantage was far more difficult than many thought. We believed that much more attention should be paid to the pragmatics of strategic planning per se—with or without information technology. We believed that the identification and implementation of strategic systems is much harder than appreciated. And we believed more effort should be expended in choosing, and tracking, the right technologies.

The book is divided into four sections. Section 1 describes briefly the background literature in the strategic uses of information technology and organizational planning. Section 2 considers the theoretical issues involved in thinking about strategy and organizational process. Section 3 focuses entirely on implementation of strategic systems. And Section 4 looks ahead to new and promising technologies which may play a part in future strategic systems.

We want to thank our colleagues who participated in the Symposium by writing papers and providing criticism. The NYU Symposium Series on Information Technology provides us with a unique public forum to raise issues, and we express our thanks to our NYU colleagues in the Information Systems Area for their support and encouragement. Many IS Department graduate students helped in recording papers—and question and answer sessions which followed the papers—for which we are grateful. The staff of the IS Area, particularly Carole Larson, provided patience as well as support for this entire project.

Ken Laudon
Jon Turner

INFORMATION TECHNOLOGY AND MANAGEMENT STRATEGY

INTRODUCTION

SECTION 1

If one theme has captured the imagination of business writers in the last several years it is the way that *information technology* (IT) has been used as a competitive weapon and even to restructure industries. By now, almost everyone is familiar with American Hospital Supply, the company that installed terminals in their clients' offices, permitting AHS to provide better service in supplying medical goods, at a reduced cost, and at the same time, locking in their customer base. AHS's position as the dominant firm in their industry is attributed to this strategy.

Does it really happen this easily? Do executives scan the environment looking for opportunities to apply information technology? Is top management really tracking technology? Are firms likely to be any better at capitalizing on the use of IT to achieve strategic objectives than they are at building normal information systems, or for that matter, accomplishing any strategic goal?

In this section we first trace the evolution of the notion of using information technology to accomplish strategic objectives. Next, the literature on organizations and strategy is examined to identify critical questions involving the use of IT. Finally, the major themes of the book are introduced.

STRATEGIC USE OF INFORMATION TECHNOLOGY

Gorry and Scott Morton (1971) were among the first to recognize distinctions among information systems. They arrayed them by managerial activity* and the types of decisions involved, pointing out that the areas of most concern to management were unstructured decisions. They argued that the greatest payoff to organizations occurs at a strategic level, where problems are rich in ambiguity and uncertainty. Decisions at this level are highly leveraged—improvements could have a high payoff for things that really matter. This work led to the recognition that decision support systems (DSS) were distinct from normal data processing systems. Gorry and Scott Morton's focus, however, was on using computer technology to improve managerial decision making through a process of model building and learning. If a competitive advantage occurred as a result, it was coincidental.

In 1982, Hank Lucas and Jon Turner observed that information technology can be used to achieve strategic managerial objectives in three fundamentally different ways (Lucas and Turner, 1982). First, it can be used to obtain greater *efficiencies* in existing operations: for example, in reducing variable costs (decreasing marginal costs) through automation of routine activities, or in improving services to clients through better use of information. It can also be used to improve the *strategic planning process* by the development of policy analysis support systems. Finally, technology can be used to open *new markets* through the development of novel products or services that rely on or incorporate technology directly. Although the authors identified many of the factors involved in executing these strategies, the focus was still on technology, stressing (or classifying) the way it was used.

Greg Parsons noted a year later that if information technology was to become a feasible competitive weapon, senior management would have to understand how its use affected the competitive environment and the strategy of a firm (Parsons, 1983). He identified three levels of interaction. At the *industry* level, IT can change the products and services offered, the markets, and the economics of production through fundamental changes in processes. At the *firm* level, IT affects relationships among buyers and suppliers, influences substitution, acts as a barrier to new entrants, and is an arena in which rivalries can be contested. Finally, at the *strategy* level, a firm can use IT to become a low-cost leader, to differentiate its products from others in the field (usually by incorporating technology into them, for example as a control element or to provide a display), or to customize the product so that it satisfies a market or product niche. The significance of Parsons' work was that it shifted the analytic perspective away from technology to a firm's strategy, and incorporated the vocabulary of strategic planning.

*Strategic planning, management control, and operational control were the categories used to represent managerial activity, following on work by Anthony.

Warren McFarlan (1984) expanded on this theme by observing that information technology can add value to products and be used to throw competitors off balance. The issue, from his perspective, is *searching for opportunities*. IT can be used to build barriers to entry. IT can change the balance of power in supplier relationships and generate new products. IT can be used to build in switching costs and to change the basis of competition. Here McFarlan, as Parsons had done, directly links technology to Porter's three categories of competitive strategy: cost based, product differentiation, and market niche (Porter, 1979).

Porter clarifies many of these issues in his paper on how information technology can be used to obtain a competitive advantage (Porter and Millar, 1985). Information technology is conceived even more broadly as *information* that a business creates and uses as well as the wide spectrum of increasingly convergent *technologies* that process this information. Porter uses the notion of a "value chain" to describe ways in which IT can be used to reshape the process by which companies create their products as well as the products themselves. An important concept here is that of "linkages," where the way in which *one* activity is performed affects the cost or effectiveness of *other* activities. By managing these linkages, a firm can gain advantages, often by passing costs on to others, who come either earlier or later in the value chain, such as suppliers or distributors, or through fundamental alteration in the process of production. Porter insightfully observes that every value activity has both a *physical* and an *informational* component. Although the tendency has been to focus more on management of the physical activity, information and processing associated with a value activity have many unexplored opportunities for advantage—sometimes even being sold as a separate product.

Porter's focus on information and how it can be used constructively to advantage is an important contribution. Not only does it shift the major interest away from technology itself, but it ties the thrust back to original decision-making themes. It is, of course, the information provided by technology that is important rather than the technology itself.*

Williamson (1975) suggests that there are two basic ways of coordinating organizational activity. *Internal hierarchies* can be used to establish price, quantity, quality, and schedule for a given product. In this case suppliers are not selected from a group of potential candidates; they are directed by management. Conversely, an *external market* can be used to determine suppliers based on their ability to meet cost, quantity, quality, and schedule objectives. This approach uses traditional market mechanisms of supply and demand to provide information for selection and coordination.

Relatively little information is needed to accomplish coordination with hierarchies because choices are limited and predetermined. In this case

---

*However, as Lucas and Turner (1982) observe, this presumes that an organization has control over its technology and can shape it to achieve desired objectives.

production costs tend to be higher without competition to drive them down. With external markets, considerable information must be gathered and processed to identify appropriate suppliers (and control them). As a result of the competitive market mechanism, however, production costs will be less. Thus a trade-off exists for a firm between costs of coordination that are primarily information processing, and costs of production (Malone et al., 1986).

In summary, the field has moved away from classifying uses of information technology toward an inspection of all aspects of a company in search of opportunities to apply information to gain a competitive advantage. Although this is an impressive shift in perspective, it does raise some difficult questions given our current understanding of how organizations function.

AN ORGANIZATIONAL PERSPECTIVE

Literature concerned with the strategic role of information technology has largely ignored the organizational context in which *all* strategic planning must be done. Although it is valuable to recognize the potential role that information and information technology may play in organizational life, and it is useful to study selected cases where IT has had a powerful strategic impact, fundamental questions remain: How do organizations conceive of and implement strategic plans of any sort? What do we really know about strategic planning in organizations?

Virtually all of the literature on the strategic role of information technology cited above assumes that organizations are rational goal-seeking entities composed of more-or-less like-minded persons who share goals and who work cooperatively toward common aims. Internal environments, in this view, are largely supportive. When they change, organizations adapt by anticipating them and then carrying out change programs (Zaltman et al., 1973).

This venerable and sanguine view of organizational life tends to overlook some powerful findings of organizational research in the last 25 years. First, organizational action is in part determined by the outcome of political struggles among factions and personalities within organization (Allison, 1971). The ability of an organization to perceive a changing reality correctly and to adopt "the right" strategic plan or technology depends on the outcome of these clashes (for a review, see Laudon and Laudon, 1988).

Second, organizational action, including planned strategic use of IT, is largely constrained by bureaucratic routines honed over many years of practice. These routines do not change rapidly (Hannan and Freeman, 1977). Organizations tend to respond to changing environments (internal or external) by doing more of the same thing as long as possible until, often, it is too late.

A strategic plan, then, tends to reflect not so much what the organization should do under ideal conditions, but rather what the organization is prepared to do given the kinds of trained specialists, bureaucratic units, and routines available.

Third, the perceptions that organizations and their leaders have of the environment are heavily influenced by powerful, unstated cultural assumptions about the nature of the business, employees, products, customers, and suppliers (Schein, 1985). The ability of organizations to perceive the environment correctly and to adapt rationally by forming reasonable strategic plans is therefore circumscribed by these bedrock cultural assumptions.

Finally, there is an important sense in which organizational life is random, or subject to dynamics and developments that appear random. Issues and problems arrive in organizations in an unplanned, unpredictable sequence. Solution producers—all highly trained specialists—are always searching for new areas to practice their preferred brand of solution (Cohen et al., 1972). Time-driven events and routines activate various decision makers and force action. Decision makers often attach solutions to problems in a haphazard manner, resulting in a series of uncoordinated movements (Starbuck, 1983).

The view of organizations that emerges from this body of research is that what organizations actually do is often unreflective of their environment and often not the right thing at the right time (see Laudon, 1985). Most organizations fail, and very few of them live longer than a human lifetime (Hannan and Freeman, 1977; Starbuck, 1983). What organizations plan to do (the formal strategic plan) is usually very different from what they actually do (the emergent plan) (Mintzberg and McHugh, 1985). Virtually all empirical research on strategic planning has found that organizations that engage in strategic planning do no better (in any measurable sense of the term "better") than firms that do not practice it (Grinyer and Norburn, 1975).*

On the other hand, this brief review of organizational research raises some troubling questions for researchers and managers engaged in the strategic use of information technology. First, how can organizations discover appropriate strategic opportunities, some of them involving the use of information technology? Second, how can organizations develop the necessary political and bureaucratic support to implement strategic IT investments? Third, how well does the strategic planning cycle for IT fit with that of the firm? In other words, how does the input, process, time line, and output of IT strategic planning mesh with that of the normal planning process? Is it possible, or even desirable, to coordinate strategic IT investments closely with the

---

*This may all turn out to be a great blessing. If organizations actually carried out their five-year strategic plans, they would have to ignore important environmental changes which occur in shorter time frames.

plans of the firm? Finally, where in the normal strategic planning process does consideration of the opportunities presented by IT take place, and what is necessary for this to occur? It is these issues that are addressed by the authors in this book.

RESEARCH ISSUES

The field has moved from a perspective of classifying the strategic uses of information technology to describing its successful strategic application. It is time, however, to go further and ask deeper research questions about the process by which these outcomes come about.

There is a tendency to think of technology as the driving force behind strategic applications of IT. The notion set forth in the literature is that decision makers sit around inspecting the value chain and linkages searching for opportunities to apply IT in order to gain competitive advantage. This view may place undue emphasis on IT, and certainly it is top down. In actual fact, IT may well play much more of a secondary role. The strategic application may well be the result of a keen analysis of business needs. It could also be that really good IT applications come from the bottom up, fighting to overcome social inertia, gathering political and popular support along the way—a thought consistent with studies of other innovations. Consequently, it is important to learn much more about how strategic applications were *conceived*. One example of a long term study is that of Laudon and Westin (1989) which traced systems development at the Social Security Administration from 1935 to 1990.

Second, given the results of organizational research into strategy formulation, much more careful analysis is needed of just how strategic IT plans were, in fact, *implemented*. There is a general lack of attention to this process in the existing literature.

Third, much more information is needed about *failed* strategic applications of IT. There have indeed been some colossal ones: Federal Express's ZAP MAIL, IBM's and Merrill Lynch's stock quotation system, and the IBM-Aetna Satellite Business Systems venture, to mention a few. It may be much easier to identify the causes of failure than to agree on the factors that lead to success.*

Finally, the implicit model behind much of the literature is that building strategic applications is an example of purposeful, goal-directed behavior.

*The research approach, currently in vogue, of describing successful applications of IT in case format is misleading because it overemphasizes successful outcomes. Cases, which tend to be written after the fact, aggrandize successful results, rationalize behaviors, and infrequently capture tensions and negative events. As such, they can present a distorted and misleading view.

There is some indication that the strategic use of information technology has an opportunistic component. Projects started for one purpose often end up being used for a totally different one. It is not at all clear that American Airlines built Sabre with the notion of controlling the flow of information to travel agents, thereby promoting their own flights over the competition. More likely, they decided to replace some of the manual record keeping supporting their counter staff for purposes of improved accuracy and timeliness and then, slowly, began to realize the opportunities the system presented for being used in a competitive (strategic) manner.

Sabre probably was not built because deregulation was anticipated, but it did permit American to have the information needed so that they could experiment with many different fare structures and optimize their return. American Hospital Supply—often touted as the premiere example of the planned use of strategic systems—actually evolved a number of systems slowly over a long period. In reality, strategic uses of information technology may require supportive cultures to thrive, may have to be nurtured for quite a while, and may need heavy commitments of human and financial resources.

PLAN OF THE BOOK

The articles in Section II describe the organizational context or setting in which strategic moves are made. Section III contains articles that illustrate specific implementations of strategic systems. Section IV covers important underlying technologies and shows how they relate to the process of strategy formulation.

REFERENCES

ALLISON, G. T. (1971). *Essence of Decision,* Little, Brown, Boston.

COHEN, M. D., J. G. MARCH, and M. D. OLSON (1972). A garbage can model of organizational choice, *Administrative Science Quarterly* 17.

GORRY, G. A., and M. S. SCOTT MORTON (1971). A framework for management information systems, *Sloan Management Review* 13(1):55–70.

GRINYER, P. H., and D. NORBURN (1975). Planning for existing markets: perceptions of executives and financial performance, *Journal of the Royal Statistical Society, Part I* 138.

HANNAN, M. T., and J. FREEMAN (1977). The population ecology of organizations, *American Journal of Sociology* 82.

LAUDON, KENNETH C. "Institutional and Environmental Models of System Development," Communications of the ACM, July 1985.

LAUDON, KENNETH C. and LAUDON, JANE P. Management Information Systems: A Contemporary Perspective. New York: Macmillan, 1988.

LAUDON, KENNETH C. [With Alan F. Westin], Systems Development at Social Security: 1935–1990. Forthcoming, 1989.

LUCAS, H. C., JR., and J. A. TURNER (1982). A corporate strategy for the control of information processing, *Sloan Management Review* 24(3):25–36.

MALONE, T. W., J. YATES, and R. I. BENJAMIN (1986). Electronic markets and electronic hierarchies, *Proceedings of the 7th International Conference on Information Systems*, pp. 109–112; forthcoming in CACM.

MCFARLAN, F. W. (1984). Information technology changes the way you compete, *Harvard Business Review* (May–June):98–103.

MINTZBERG, H., and A. MCHUGH (1985). Strategy formulation in an adhocracy, *Administrative Science Quarterly* (June).

PARSONS, G. L. (1983). Information technology: a new competitive weapon, *Sloan Management Review* 25(1):3–14.

PORTER, M. E. (1979). How competitive forces shape strategy, *Harvard Business Review* (March–April):137.

PORTER, M. E., and V. E. MILLAR (1985). How information gives you competitive advantage, *Harvard Business Review* (July–August):149–160.

SCHEIN, E. H. (1985). *Organizational Culture and Leadership*, Jossey-Bass, San Francisco.

STARBUCK, W. (1983). Organizations as action generators, *American Sociological Review* 48.

WILLIAMSON, O. (1975). *Markets and Hierarchies*, Free Press, New York.

ZALTMAN, G., R. DUNCAN, and J. HOLBEK (1973). *Innovations and Organizations*, Wiley, New York.

THE ORGANIZATIONAL CONTEXT

SECTION 2

Articles in this section seek to describe realistic organizational settings within which all strategic uses of information technology must take place.

William Starbuck argues that most organizations do not successfully carry out their strategic plans, as indicated by the enormous failure rate of organizations. The realistic potential for using information technology to gain long-term strategic advantage is much more limited and problematic than previous literature based on glowing consultant "experience" would suggest.

The problem, according to Starbuck, is not so much that environments are more complex, or changing more rapidly, than before. Rather, it is that organizations build in failure as they attain success. By virtue of learning to do a few things very well, they ossify and become rigid and brittle. To break the stranglehold that past success can exercise over corporate life, it is necessary to replace the top management group.

Vitale, Ives, and Beath extend the critical assessment of Starbuck by focusing specifically on efforts to use information technology for strategic advantage. They are critical of the Pollyanna quality of the current literature, much of which rests on a top-down model of strategic systems development in which practitioners are told to align information technology with the strategic goals of the corporation. This top-down model simply does not work in most organizations because an overall strategy does not exist. Hence there is nothing with which to align information technology. Second, organizational strategists frequently do not understand the technology. Then environmental turbulence assures that systems developed for one environment are quickly out of date. Vitale, Ives, and Beath go on to explore an adaptive strategy model as potentially more useful for developing strategic IS applications. This adaptive strategy is especially sensitive to the implementation hurdles faced by an new system.

King, Grover, and Hufnagel extend the critical understanding of the literature further by clarifying the difference between *strategic uses of information* as opposed to strategic uses of *information technology.* Often, unfortunately, organizations fail to make this distinction. Because information technology is more tangible, organizations tend to focus on the hardware and miss strategic opportunities based on new, less tangible uses of information. Equally important for King et al. is the need for greater precision in defining the *strategic* importance of systems. In many instances, previous authors have failed to distinguish truly *strategic systems* from merely *important systems.*

King et al. conclude with a survey of IS executives seeking to identify the most common areas of strategic application and the organizational factors that facilitate or inhibit strategic system development. As it turns out, in most organizations, the development of strategic systems is haphazard.

CHAPTER 1

Why Organizations Run into Crises ...and Sometimes Survive Them

William H. Starbuck
New York University

The world is a representation of our language categories, not vice versa.—Edmund R. Leach (1964)

In this paper nothing is said about computer technology. Rather, we discuss the strategies of organizations, how they get formulated, and what happens to them over time. More specifically, we focus on strategic crises, situations in which organizations confront very serious threats to their existence. We report what our studies reveal about why crises occur and what happens during crises.

From 1940 to 1975, one basic theme of American society has been economic expansion. Americans have assumed that the gross national product grows larger each year, that business firms get bigger every year, and that most people get richer every year. Even in the late 1950s and early 1960s when we were facing the threat of nuclear holocaust, Americans nevertheless thought that their economic well-being was improving and that firms were growing normally.

Possibly as a consequence of this long period of economic growth, the professional literature of managers and management professors has generally portrayed the growth of an organization as both desirable and likely, if not inevitable. Management textbooks often describe a sequence of stages that an organization supposedly passes through as it grows, and managers commonly tell each other that an organization must grow or die.

In contrast to this idyllic vision of prevalent, hard-to-avoid growth, the statistical realities are that very, very few organizations start out small and then continue growing until they become very large. Nearly all small organizations simply go out of existence within a few years without ever growing even to middle size. The great majority of middle-sized organizations start out middle-sized, and not many middle-sized organizations grow large. And many large organizations are created large. Some of the highest death rates occur among the largest new organizations (Starbuck and Nystrom, 1981).

Very high death rates also occur among the smallest organizations. Organizations that employ fewer than 100 people generally have short lives and low profits. In fact, organizations with fewer than 100 employees differ so much from organizations with more than 200 employees that they comprise a distinctly different population.

There is no statistical evidence, however, that among organizations employing more than 200 people, larger organizations have any advantage over smaller organizations. That is, the largest organizations do not survive longer than those with 200 to 1000 employees. Also, the largest organizations are just as likely to lose a certain percentage of sales, or to make a profit of a certain percentage of sales, as are much smaller organizations, with 200 to 1000 employees.

Figure 1-1 shows that almost all corporations have very short lifetimes. Business failures occur more often during recessions than during booms, of course, so Figure 1 is a synthesized curve indicating what the survival rates would be if every year were an average year. About half of all newly created corporations disappear within three years, and three-fourths within seven years.

Lasting longer improves organizations' survival chances. Figure 1-2 shows this improvement, stated as the probability of lasting an additional five years. The curve resembles a learning curve. Corporations' survival chances go up quite noticeably through the first 20 years of their existence. A brand new corporation has only about a 34 percent chance of making it for five years. But after a corporation has been in business for 20 years, it has something like a 73 percent chance of surviving an additional five years. Evidently, the surviving organizations translate their experiences into fewer mistakes and strong bonds with their environments. Of course, less than 8 percent of the brand new corporations make it through 20 years. There really are no data for periods beyond 25 years, but the curve does seem to continue up and up toward 100 percent; 350-year-old Harvard University probably has a very high likelihood of still being in business five years from now.

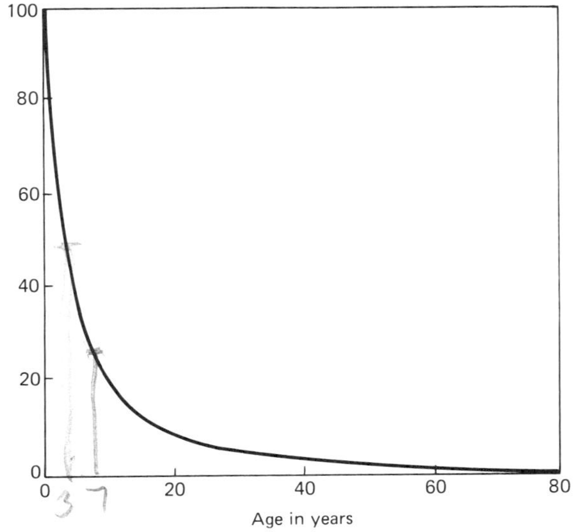

Figure 1-1 Percentages of corporations surviving to various ages.

One reason that organizations find survival so difficult is that improving organizations does not normally make them better able to survive. If somebody invents a superior managerial practice—such as management-by-objectives, or cost accounting, or electronic spreadsheets—and if the inventors use this practice in just one organization, that organization might operate more effectively. But if other organizations take note of this new prac-

Figure 1-2 Percentages of corporations surviving five years after various ages.

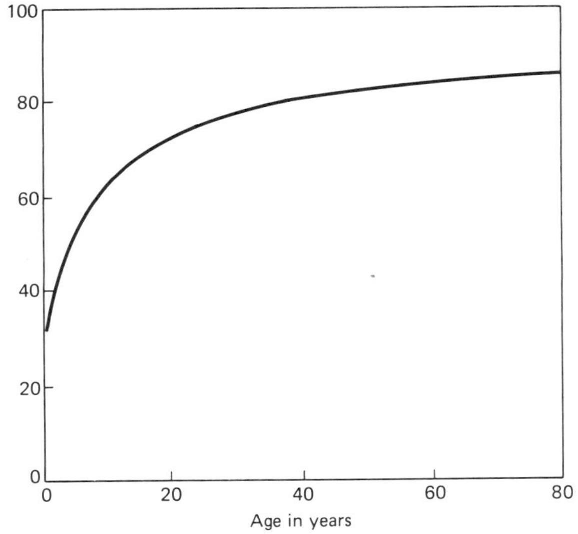

tice and adopt it, so that it spreads generally through the whole population of organizations, none of the organizations reaps benefits individually. Consumers might find themselves better off, or employees might fare better, or society at large might benefit from less waste or less pollution; but the adopting organizations will not last any longer or earn higher profits, because the adopting organizations will not have comparative advantages with respect to each other. all alike

Comparative advantages are not easy to come by, and are nearly impossible to keep. Large organizations especially have trouble getting and keeping comparative advantages. My experiences as a consultant suggest that the most profitable business firms are rather small ones. A small firm does not draw much attention, so it can exploit a favorable niche and earn very high profits without attracting competitors.

DO ORGANIZATIONS OR THEIR ENVIRONMENTS CAUSE CRISES?

The foregoing vision describes where my colleagues and I ended up after about 8 or 10 years of research. Our research actually started from quite a different viewpoint, and we did not even think of looking at statistics at the start. I was living in Berlin and I had a Swedish colleague, Bo Hedberg. Bo had gone home to Sweden, and when he came back to Berlin, he said he had received a research grant from the Swedish government to study stagnating industries. He said that he wanted to find out why some industries stagnate and drive firms out of business and put people out of work.

My background and orientation are social–psychological, so I tended to place the responsibility on the firms rather than their environments. I said something like: "Bo, you've got this thing all screwed up! The interesting question is not why some industries stagnate. Technologies are always developing, populations are always migrating, prices are always shifting—it is inevitable that things will change, and some of these changes will render some industries obsolete. The relevant question is: Why does an intelligent group of people who are running a firm choose to remain in a stagnating industry even though they recognize that it is stagnating? Why don't firms move into more promising industries when their current ones start to stagnate?"

Bo responded, approximately: "You're being unrealistic. Somebody who knows how to make and sell something can't just pick up their product line and their engineers and plunge into another industry. Their specialized skills and business connections make them captives of their environments. The firms in an industry have to evolve together. It is a societal problem to create incentives that keep industries vital, that keep them evolving in line with social needs and economic and technological opportunities."

Obviously, we had an argument: I was saying that industrial stagnation posed problems for the managers of individual firms, whereas Bo was saying that industrial stagnation posed problems for the setters of governmental policies. We decided to resolve our argument by doing research together.

One result was a dozen case studies of organizations facing crises. We chose to make case studies because we thought we were looking for rare and unusual events. We believed that nearly all organizations and industries go along happy, healthy, wealthy, and wise most of the time; but that once in a while, a disease, much like a virus or a bacterium, infects an organization or industry and makes it ill. When the disease is very serious, the afflicted organization or industry suddenly faces a crisis that it cannot survive, and it goes out of business. We saw ourselves as organizational pathologists. A pathologist is a doctor who cuts up a dead body to see what it died of. We went looking for organizations that were facing very serious crises so that we could see what caused them to get into so much trouble and what they did once they were in it.

I will use one of our cases as an illustration. It involves a large, well-known company, and it makes a good illustration because it incorporates most of the elements we observed in other cases.

Facit AB is a Swedish company that was founded in 1920 to manufacture mechanical calculators. It operated with great success for about 50 years. Besides mechanical calculators, it made typewriters and office equipment such as desks and chairs, and a few of its personnel had dreamed of someday making electronic equipment. In the mid-1960s, Facit borrowed large sums in order to build new factories; and by the late 1960s, Facit's personnel were taking considerable pride that they could build better mechanical calculators at lower costs per unit than any other company in the world. By 1969, Facit operated 20 factories in five countries and sales offices in 15 countries, and it employed 14,000 people. Employment had gone up by 70 percent and sales and profits had doubled between 1962 and 1969 alone. By everyone's standards, Facit was an extremely successful company.

1969 was a peak year. Suddenly, electronic calculators appeared on Facit's horizon. The immediate reaction of Facit's senior executives was that these funny little plastic boxes with little red lights on them had to be a passing fad. It was obvious to Facit's senior executives that no one would trade in a nice solid machine that went thumpity, thumpity, thump for a silly little box with little red lights on it! Facit would move into electronics gradually over the next decade or two, but it would do so in a careful, orderly way that maintained high product quality.

Three years of layoffs, declining sales, and negative profits followed. The price of Facit's stock dropped precipitously. Several factories were closed;

these were typewriter and office-equipment factories because Facit's top managers, knowing that eventually mechanical calculators would come back into favor with customers, were raising cash in order to subsidize the sale of mechanical calculators. The only problem was to retain loyal customers until the demand for mechanical calculators recovered.

A new managing director was appointed several times, roughly every six months (the managing director is the chief executive). Each new managing director announced that his predecessor had misrepresented the company's situation, which was actually much worse than his predecessor had publicly admitted; but the new management team knew what really needed to be done, so the future looked rosy.

Meanwhile, the losses continued. In 1971, only one division of Facit made a profit—the typewriter division. In February 1972, the head of this division quit in protest because his colleagues were planning to close yet more typewriter factories.

By the summer of 1972, employment was down to 9000, and Facit's senior executives had reached a state of utter despair. They realized that they did not know what they should do or how they could save their company. Whereas they had assumed at first that Facit could shift to electronics gradually, it now appeared that a technological revolution had passed them by and there was no way that they could ever catch up with it. Facit was a mechanical-calculator company in an electronic-calculator age.

They hired McKinsey & Company to tell them what to do. McKinsey basically agreed with the depressing analysis of Facit's senior executives, and recommended that Facit fire an additional 2400 employees and close yet another plant. This happened to be a large plant in their small corporate-headquarters town in Sweden, a closing that would have disastrous effects on the town.

In early October 1972, Facit's board of directors met for three days and nights to discuss McKinsey's report, but they could not reach a decision. On the one hand, they saw the sense of what McKinsey was recommending: the only hope for the company was to retrench, and it needed cash to meet immediate obligations. On the other hand, destroying your hometown is a very, very difficult thing to do. Unable to reach a decision, Facit's board contacted a senior executive in another company. They asked him to come in as their newest managing director and to make this decision. He said he would have nothing to do with it!

That is exactly the kind of story for which Bo and I looked. Together with Paul Nystrom, we studied specific cases and tried to draw inferences from them. We think we learned some interesting lessons, but I must warn you how difficult it is to draw meaningful inferences from empirical observations. Researchers bring biases to their observations and interpretations of events; and events that happen spontaneously can be explained in many different ways.

I learned this lesson from a story told by Kenneth Pelletier, a doctor from

Berkeley, California. Pelletier wanted to investigate how it is that people live to be 140 years old, so he applied for a federal government grant for this purpose. To generate hypotheses for his grant application, he identified four societies in which it was supposedly quite common for people to live to 140 years of age, and then he looked for the common properties of these societies.

He could quickly rule out climate because the societies are in very different climates. He could also rule out medical care, because none of them had much of that. But he did spot five properties shared by the four societies.

First, they are vegetarian, although not necessarily by choice. They have little access to meat.

Second, in all of these societies, everyone works hard throughout their lives, and no one can retire. If people quit working, they quit eating.

Third, these societies exhibit a great deal of social solidarity. If a man and his wife live together in a hut at the edge of a village and one of them dies, the survivor is taken in by other folk in the village and made a part of their family.

Fourth, about once a week in these societies, everybody in a village gets together and gets blind, roaring drunk. Pelletier speculated that this alleviates stress and fosters solidarity, while the social context and periodicity prevent alcoholism.

Finally, people in these societies continue to have sexual intercourse at least to the age of 120. I suggest that this condition is sufficient to explain the phenomenon!

LEARNING TO FAIL WHILE LEARNING TO SUCCEED

One risk is that research will mainly reinforce beliefs the researchers held before they started. Bo, Paul, and I certainly brought prejudices to our studies, but what we found out was not what we had initially expected to find out. Perhaps, this is because we had initially disagreed with each other, so we forced each other to consider alternative interpretations of events. Indeed, several of the key lessons we drew only came to us after some years of study and over our resistance.

For example, we began with the idea that we were looking for abnormality—a rare disease that happens to infect a few organizations or industries. But we were inundated with examples. Crises appeared wherever we looked. Indeed, we did not even have to look for them. They infested our newspapers and magazines; our colleagues pointed them out to us; skeptics in our audiences brought us examples. It gradually dawned on us that we were seeing normality: Failures and real threats of them are common. Every organization confronts a crisis at some point, and some organizations survive several crises. Thus, once we started looking for crises, we had no trouble seeing them.

Furthermore, we had little difficulty obtaining information about what happened before and during crises. Earlier in this conference, someone asserted that people do not talk about failures. Our experience was that people are very eager to talk about major dramas in their lives, such as threats to their continued employment, or their struggles against and triumphs over adversity. People are even happy to talk about failures, their own failures, but especially those of their colleagues.

It took us years of struggling with these phenomena to see that one reason crises are so prevalent is that failures arise from precisely the same causes as do successes. Crises result from too much of a good thing, in the sense that the very processes that enable organizations to achieve success also lead them into error.

Consider the means by which organizations learn. Most powerful tools are two-edged swords that can produce harm as well as good. Organizational learning mechanisms show this duality: while they foster autonomy, efficiency, and predictability, they simultaneously promote blindness, rigidity, and self-deception.

Organizations have three basic learning mechanisms. One of these is buffering. An organization builds buffers between itself and sources of random variation in its environments. A good example is a finished goods inventory. This inventory allows an organization to run its production processes at stable rates, pretty much independent of the fluctuations from day to day in customers' orders. The inventory will soak up those fluctuations provided that the organization attains an approximate, long-run match between the rates of production of its various products and their rates of sales. Buffers embody knowledge about the ranges of variation in environmental demands and how to satisfy environmental demands at low cost to the organization itself.

Slack resources afford a second learning mechanism. Managers hide slack resources in overly generous reserves or behind unnecessary expenditures—expenditures that an organization does not really have to make, such as new curtains or new carpets or walnut paneling in offices. Such expenditures may yield some short-run benefits—they may make employees more comfortable, for example—but if times call for stringent economies, it is easy to cut back on those expenditures and thereby to gain access to additional resources. Slack resources resemble buffers in that they enable an organization to satisfy environmental demands at low cost; and by lowering what would otherwise be peak profits, slack resources also make an organization's performance appear smoother to its environment.

The third, and most important, learning mechanism is programming. Programs enable organizations to repeat the same activities over and over again. People figure out how to solve some kind of problem or how to perform some task effectively, so they create a program that enables them to do it over and over again the same way. Most organizations, in fact, write many programs down as standard operating procedures; a lot of big companies hand their new employees books filled with procedures they are supposed to follow. One man who was the industrial-relations manager at his plant told me that he had made a point of becoming the keeper of the procedure manuals because this gave him considerable power: He knew what everyone was supposed to do and how they were supposed to do it. Programs afford the main means by which organizations accumulate experience, coordinate activities, and control actions hierarchically.

I am especially intrigued by programs of a kind that I call *action generators*. Action generators are activated by things like calendars and clocks and job descriptions rather than by information-bearing stimuli. People may get up in the morning and go to work and carry out certain job responsibilities even if nobody else has need for their help on that day. An accountant compiles a monthly report, not because someone explicitly requests it, but because the accountant compiles this report at the same time every month; the accountant may have no assurance that anyone is actually going to read the report this month. A quality-control inspector pulls a sample and conducts a test even if no explicit question about product quality has arisen, because it is the inspector's assignment to monitor product quality and the organization has set up procedures for testing quality and it has provided a budget and floor space for quality testing. Similarly, organizations fabricate products, produce budgets, conduct research, and advertise without regard for whether these actions can solve visible problems or meet immediate needs.

Most organizations make prolific use of action generators and assign them important functions; but more successful organizations probably depend more strongly on action generators, because successes breed complacency, whereas failures and difficulties encourage reflection.

These learning mechanisms—buffers, slack resources, and programs—offer many advantages: they preserve some of the fruits of success, and they make success more likely in the future. They stabilize behaviors and enable organizations to operate to a great extent on the basis of habits and expectations instead of analyses and communications. They reduce the complexity of social relations and keep people from disobeying or behaving unpredictably. They minimize needs to communicate or to reflect, and they conserve analytic resources. They also give organizations discretion and autonomy with respect to their environments. Organizations do not have to pay very close attention to many of the demands currently arising from their environments, and they do not have to formulate explicit or unique responses to most of these demands. Thus organizations gain human resources that they can devote to influencing their environments and creating conditions that will sustain their successes in the future.

But these learning mechanisms also carry disadvantages. In fact, each of the advantages has a harmful aspect. People who are acting on the basis of habits and obedience are not reflecting on the assumptions underlying their actions. People who are behaving simply and predictably are not improving their behaviors or validating their behaviors' appropriateness. Organizations that do not pay careful attention to their environments' immediate demands tend to lose track of what is going on in those environments. Organizations that have discretion and autonomy with respect to their environments tend not to adapt to environmental changes; and successful organizations want to keep their worlds as they are, so they try to stop social and technological changes. Indeed, buffers, slack resources, and programs make stable behaviors, current strategies, and existing policies appear realistic by keeping people from seeing problems, threats, or opportunities that would justify changes.

The organizations we studied, and probably most organizations, develop perceptions of themselves and their environments that diverge greatly from what an outsider might regard as objective reality. In extreme cases, organizations' perceptions become as unrealistic as the perceptions of schizophrenic people.

Consider, for example, some of the perceptions of Facit's senior executives. The senior executives defined their company solely as a mechanical-calculator company. They did not define it as an office-equipment company in a broad sense, nor even as a calculator company in a broad sense. Because of this very narrow and specific definition of their company's product line, they thought of typewriters and office equipment as peripheral by-products of their core technology, mechanical calculators; and they thought of electronic engineering as peripheral expertise, largely irrelevant to their product line. Faced with a serious threat, they decided that they had to protect their core technology; and to obtain funds to do this, they felt that they could sell off the peripheral activities, typewriters and office equipment. An outsider might have remarked that mechanical calculators constituted only a third of Facit's sales volume before electronic calculators appeared, and that electronic calculators did not directly threaten Facit's markets for typewriters and office equipment. An outsider might also have viewed electronic calculators less as a threat than as a natural and appropriate addition to Facit's product line. Indeed, an outsider might have noted that back in the 1960s, Facit had started a joint venture with the Sharp Corporation of Japan, and this subsidiary had designed electronic calculators and computer terminals. During the late 1960s, this subsidiary had proposed that Facit start making and selling these products, but Facit's board of directors had rejected this proposal. So Facit actually had access to a whole set of electronic products throughout its crisis.

Facit's senior executives also looked upon technological change in their industry as a slow, incremental process. This fitted their experience, for mechanical-calculator technology had been evolving in a linear, incremental way for 50 years. The gears had gotten a bit smaller each year, the motors had gotten a little bit faster each year, and the boxes had gotten a little bit smaller each year—but product changes had been small improvements that could be extrapolated from history. Because they had not been monitoring developments in electronic technology closely, the advent of electronic calculators took

Facit's senior executives largely by surprise. And because they were accustomed to slow technological change, their perceptual frameworks gave them no bases for dealing with the sudden emergence of a totally different technology. Their first reaction to electronic calculators was to fall back on their experience and to assume that electronics would enter their product line in a slow, incremental way.

Two years later, Facit's senior executives concluded that the world had passed them by. They saw Facit as a ponderous machine that could only change slowly and incrementally, yet it was now apparent that Facit had to compete with a rapidly evolving new technology for which it had no capability. Facit could not move into electronics as quickly as market pressures demanded. This second conclusion was just as unrealistic as the first had been—as you will see.

ARE ORGANIZATIONS OR THEIR ENVIRONMENTS TO BLAME?

Bo and I eventually resolved our argument. Fortunately for our friendship, we resolved it by concluding that both of us had been partly right. Of course, such an outcome may also have been inevitable consequence of our friendship, which had grown much stronger over the years of cooperation.

Crises are indeed produced by organizations' environments, although not exactly in the way that Bo had initially conceived. Bo had rightly seen that environments change so as to obsolete some markets, products, and technologies, but he had not seen that environments also make it hard for organizations to adapt to these changes. Environments propound a set of ideas about how to organize, and then the environments behave in ways that make these ideas unrealistic. Environments tell organizations to predict the future and to develop formal strategies that fit these predictions; yet environments themselves change in unpredictable ways. Environments instruct organizations to use hierarchical control systems that make subordinates obey their superiors; yet environments make profound changes in shop-floor production technologies, consumers' tastes, and economic opportunities that are more visible to lower-level personnel than to top managers. Environments teach organizations that their decisions and strategies should be explicit and backed up by rational arguments even though explicit, rationalized decisions and strategies are harder to change than ad hoc ones.

Those who advocate long-range predictions, explicit strategies, hierarchical controls, and rational decisions might well be right if they could assure that decisions, predictions, and strategies were correct, and that top managers were up to date and realistic. But managers with distorted, obsolete perceptions may make incorrect predictions and faulty decisions, and may formulate strategies that yield very bad results. In the presence of errors and mistakes, explicating strategies and rationalizing decisions may simply make mistakes harder to correct. In the best study of formal planning to date, Grinyer and Norburn (1975) found that relatively profitable firms are about as likely to plan informally as formally, and the same is true of relatively unprofitable firms. Similarly, profitability correlates inconsistently and meaninglessly with the degrees to which top managers' agree about their firms' goals or their own responsibilities and roles. Thus, formalized plans seem to be inferior to informal plans nearly as often as they are superior, and consensus can be as harmful as beneficial when there is no way to assure that the objects of consensus are good.

Crises are also produced by organizations themselves, but somewhat differently than I had conceived at first. Although I had seen that organizations make mistakes, I had underestimated the pervasiveness of successful organizations' efforts to stabilize their environments. These efforts are logical. A successful organization knows how to succeed in its current environments, and it is not sure that it would be equally successful in altered environments, so it wants its environments to stay as they are. Unfortunately, this logical idea is not very realistic. Any single organization can do practically nothing to prevent technologies from evolving, people from moving around, or prices from changing. Our studies show that top managers, acting as a group, do have the power to stabilize activities inside their organization: they can prevent their organization from adopting new technologies or from entering new markets. But top managers have very little influence on activities occurring outside their organization, so their organization's environments go right ahead and create new markets and technologies. The top managers' stabilization efforts only undermine their organization's match to its environments and keep the organization from migrating to more fruitful environments as they develop.

Organizations also blind themselves to those environmental activities that deviate from their beliefs and expectations. Believing that certain markets, prices, and areas of expertise deserve attention, managers may assign personnel to monitor them. But no organization can afford to monitor everything, so by not assigning personnel to monitor other markets, prices, and areas of expertise, the managers are implicitly shoving these into the background (Normann, 1971). Moreover, an organization finds it much cheaper and easier to monitor environmental activities that tie into its current programs and practices.

The resulting blind spots have the interesting ability to transform beliefs into realities. For example, when Facit's senior executives wanted to assess the demand for electronic calculators, they asked their sales personnel to find out whether their customers were going to buy electronic calculators. With near unanimity, the salespeople reported back that Facit's customers said they were not going to buy electronic calculators because they preferred mechanical ones. That was probably factually a correct statement. Unfortunately, the number of customers who were able to make that statement plummeted.

Blind spots that alter realities occurred in all the cases we studied. For instance, Kalmar Verkstad had long defined its market as the Swedish railroad industry; but because this industry had lost subsidies from the government, Kalmar Verkstad was desperately searching for new sources of revenue. Through eight years of efforts to develop new products of various sorts, some of which were aimed at international markets, Kalmar Verkstad never attempted to sell railroad equipment outside Sweden, so they made no sales of railroad equipment to other countries. Similarly, Handelstidningen was a hard-pressed Swedish newspaper that was searching for cost savings. During a strategic-planning meeting, one member of the board remarked that no one bought Handelstidningen for the sports news. The board inferred from that remark that they should eliminate the sports news from the newspaper to save money. After they had eliminated the sports news, it was factually correct that no one bought Handelstidningen to read the sports news. But Handelstidningen had lost more than half of its subscribers.

Bo summarized the overall thrust of our findings with a metaphor.

> Once upon a time there was a tribe that lived in the grassland. They hunted deer, drank from the brook, and set up their tents when evening embraced them with darkness. One morning, after a gorgeous night in a pleasant campsite, some tribesmen went hunting at daybreak. They came back with deer for another day or two. Let us stay, they suggested. This is a good site to be at. So the tribe stayed one more day, and one more, and one more....
>
> Soon the hunters learned to breed their cattle, farm the land, and dam the water. As they grew wealthier and more secure, they turned their tents into houses, and then into palaces. And they fenced themselves against their enemies.
>
> The latter were known as Uncertainty, Conflict, and Ambiguity. There were enemies. Oh, yes! But the fence was sufficient. And the tribesmen bettered the defense after each attack.
>
> Pilgrims came by and told tales about better campsites and different deer. But the tribesmen paid little attention. The last hunter was dead. One summer there was less water in the brook than there used to be. No one noticed the change at first, but as the brook continued to dry out, the water manager notified the tribe council. The council fired the manager, and decided that there was water in the brook. The vote was eight to one. The tribe added to the fence, which had become so dense that one could no longer see the surrounding grassland.
>
> A few days later, the cattle were dead, suffering from heat and the

> lack of water. Looking over the fence, the tribesmen discovered that the grassland had turned into desert. They decided to move, but it was too late. Palaces have no tentpegs. (Hedberg, 1981:23)

HOW ORGANIZATIONS REACT TO CRISES

As I remarked earlier, several of our findings surprised us. One of these is that normal problem-solving methods will not lead organizations out of crises; rather, normal problem-solving methods tend to make crises worse. For example, delay is often an effective problem-solving technique. Many problems disappear or solve themselves if one only waits, or else waiting clarifies what the problems really are and what options exist. Some apparent crises also disappear after a time, but real crises grow worse as time passes, so waiting to deal with them makes them more serious. Similarly, the managers who are running an organization in a crisis cannot by themselves lead their organization out of it. For if these managers did know how to get out of the crisis, their organization would no longer be facing the crisis.

An organization that runs into a crisis goes through two or three phases of reaction—two if it does not survive, three if it does.

Weathering the Storm

The first reaction we call "weathering the storm." During this phase, a threatened organization assumes that its problems are temporary and arise in its environments. The market is adapting to a new entrant; the currency has been devalued and prices will take awhile to adjust; the temporary drop in applications is a random fluctuation. These aberrations present challenges, but surmountable ones that will correct themselves with time. Meanwhile, members of the organization must tighten their belts and intensify their efforts. But slack resources do exist that can give some breathing room. The organization can pare its budget, and it can shed peripheral, unnecessary activities. To assure sound management, the top managers will take direct control.

We have been utterly amazed at how simplistic the analyses have been in threatened organizations. Handelstidningen did not discover that it was in serious trouble until the bank called up and reported that its payroll checks were bouncing. The firm's analysis of the situation was that it lacked cash, so it got cash by selling its printing equipment. This solved its problem for only a very short period of time, because the firm had done nothing to stop the losses that were draining its funds; and having to rent printing equipment actually increased its losses.

Several of our cases involved this idea of selling assets to raise cash for short-run needs. Facit, for instance, sold its typewriter and office equipment factories. Such actions not only postpone critical appraisals of the organizations' central strategies and allow environmental pressures to build up, they also deprive organizations of resources that they could use to reorient their strategies.

Another widespread response was cooking the books—manipulating the accounting statements to make situations look better. For example, some organizations portrayed asset sales as product sales; some suspended depreciation charges; some changed their reporting periods. Ironically, these same organizations depended on their accounting reports to tell them what was wrong, and they focused on accounting data virtually to the exclusion of other information.

Three universal responses were budget trimming, shedding peripheral activities, and centralizing control. Top managers act as if they themselves are the most expert judges of what to do, and they focus their organizations' efforts on what they themselves regard as the most important activities. One consequence is that the organizations lose diversity and put more of their eggs into a few baskets. A second consequence is that entrepreneurial younger managers either get fired or depart voluntarily for organizations that offer more opportunities. A third consequence is that out-of-date people take control and push their organizations backward in time.

Top managers, by and large, turn out to be the wrong people to steer organizations out of crises—both because they cannot and because they will not. Especially in large organizations, the top managers last talked regularly to customers perhaps 20 years ago, or they last worked in factories and dealt with technologies maybe 20 years ago; and they get much of their information from secondhand reports that systematically avoid displeasing them. Top managers interpret their high positions as evidence that they have more expertise than their subordinates, and they talk and listen mainly to other top managers, who are as out of touch as themselves. Because top managers feel responsible for the policies and strategies that are now in place, they are very reluctant to admit that anything is wrong with these. They know that if their organizations' serious troubles appear to result from their policies and strategies, they are going to be made into scapegoats. Top managers also recognize that strategic reorientations would imply changes in who has control, and they do not want to lose power. So they have very strong vested interests in not admitting that anything is wrong and in not changing anything.

Consider, for example, what Facit's senior executives must have been thinking around 1970. Only a few years before electronic calculators swept into the market, the senior executives had incurred massive debt so that their company could build modern and efficient factories for making mechanical calculators, and this debt had not yet been repaid. What would their reputations be worth if it turned out that they had mortgaged their company's future to maximize its ability to produce obsolete products? And what use would Facit have for senior executives who do not understand electronics or rapid reorientations if the company shifted to electronic calculators? Facit would need different people in charge. What other companies would hire aging senior executives who had made serious strategic errors?

I thought at first that these observations about top managers were our discoveries. But then I remembered reading some similar ideas while I was attending college. Way back in 1859, Karl Marx had written that ruling social classes try to preserve their favored positions by preventing social changes. As a consequence, said Marx, technologies grow increasingly inconsistent with social structures...until the ruling classes can no longer control their societies. Then revolutions overthrow the ruling classes and transform the social structures to bring them into line with technological demands. Our case studies suggest that, for organizations, the issue is less one of technologies versus social structures than one of internal versus external changes. Top managers can block both social and technological changes within their organizations, but they exert little control over either social or technological changes in their organizations' environments. Moreover, although strategic reorientations broaden political activities and bring some new people into top management, these changes in organizations' social structures fall far short of revolutionary. Organizational hierarchies mean that large behavioral and ideological changes can result from changing just a few top managers.

Unlearning

The second reaction phase is unlearning. Unlearning seems to be more an organizational characteristic than an individual one. Individual human beings can learn without having to erase what they already know; they can record new knowledge on top of their current knowledge. This causes some confusion, because people often end up with inconsistent chunks of knowledge in their memories, but it does make learning easy. Organizations, especially the older ones, find it harder to ignore their current knowledge, because they build up explicit rationalizations for why they are doing what they are doing, and because they tend to associate specific people with specific policies. So organizations integrate their knowledge into very rigid and coherent structures in which the intellectual and political elements buttress each other. Organizations can readily learn knowledge that fits into what they already believe, but they find it very difficult to learn knowledge that conflicts with their current knowledge. Before they become willing to accept radically different knowledge, organizations actually have to unlearn what they know, by dismantling their existing ideological and political structures. In particular, before organizations are willing to contemplate radically different policies and strategies, they have to convince themselves that the arguments supporting their current policies and strategies are wrong, and they have to lose confidence in their current managerial hierarchies.

Unlearning is painful and destructive. People see their colleagues leaving because budgets have been cut, and they see entrepreneurial managers departing in search of greener pastures, so they begin to get the notion that no one with ability remains in the organization: It seems that everybody who could get a job elsewhere has done so, and that the only people left are ones who lack job alternatives or lack the ambition to pursue them. Those who remain are working harder and shouldering unfamiliar tasks, they are probably receiving less pay, and their promotion opportunities are turning sour. Slack resources no longer exist, and the mounting losses create obvious deficiencies. As resources decline, departments fight to preserve their shares or to gain bigger ones, so interdepartmental trust erodes and good faith disappears. Centralization of control makes all the more visible the top managers' ignorance of what to do: As time passes, their subordinates surmise that if the top managers really did know what they were doing, their organizations would not be in such serious trouble. Moreover, the top managers have probably declared more than once that the hard times are ending: After a while, such declarations lose their credibility and come to symbolize their speakers' impotence and unreliability.

From 1969 to 1972, Facit's managing director was replaced repeatedly and its managerial hierarchy was reorganized repeatedly. Several small factories and the main office-furnishings factory were sold. Over a third of the employees were dismissed, and continued dismissals seemed probable; the remaining employees felt lucky, guilty, and fearful. Efforts to sell a subsidiary failed. Manufacture of voting machines stopped because Facit could not produce machines of adequate quality. Every division except one was losing money. The head of the sole profitable division quit, while protesting his colleagues' policies. Local politicians sharply criticized Facit's management and urged Sweden's Minister of Industry to intervene. The news media voiced doubts that Facit would survive. Public statements by Facit's senior executives grew increasingly erratic: Claims that the company faced no truly serious problems alternated with solemn appraisals that serious problems remained ahead. In private, the senior executives came to distrust each other, and they realized that they did not know what to do.

Unlearning brings morale to a very low state—so low that many organizations simply cannot go on. Like punishment, unlearning brings an end to one line of behavior without substituting a new line of behavior, so crisis-ridden organizations discover that their old policies and strategies do not work without discovering new policies and strategies that do work. Because our first cases all ended in failures, we thought that very few organizations survive. But we eventually saw six failures and six turnarounds, so turnarounds may not be rare.

Inventing Tomorrow

Interestingly, the turnarounds we did see suggest that saving organizations from crises is both very easy and virtually impossible. Turnarounds are very easy in the sense that they require little money, little time, and little fundamental change. Turnarounds are virtually impossible because they face strong political opposition from the managers in power, who fear that they will be blamed and lose their positions.

In every one of the turnarounds we observed, some event replaced all the top managers in one fell swoop. In every one of the failures, no such event occurred; managers were replaced one at a time, but not en masse. Evidently, turnarounds require wholesale replacements of top managers.

Thus, if the chief executive of a crisis-ridden organization were to ask, "What should I do?" my response would have to be, "You should fire all your colleagues and then resign." And I would offer this advice quite sincerely...although without hope that it would be taken. The veteran top managers should be replaced even if they are all superior managers who are trying their best and even if their replacements have no more ability and less direct expertise.

It seems that a group of top managers in a crisis develops a shared vision of reality, and the only practical way to change that vision abruptly is to replace those people quickly. Under normal circumstances, a top-management group would probably encompass a range of opinions and viewpoints; but a crisis drives away dissenters and creates a sense of shared peril that homogenizes the individuals' perceptions and amplifies cohesion and conformity. A group acting together may bind themselves to their current policies and strategies even though every member of the group is individually ready for change. Bringing in one newcomer at a time does not produce significant change because the newcomer takes orientation from those with experience, and the group exerts more influence on the newcomer than the newcomer exerts on the group. A crisis means that a top-management group does not have sufficient time to reorient its vision gradually, and a crisis keeps managers too defensive and too busy fighting fires to visualize new possibilities.

A wholesale replacement of the top managers sweeps aside the strongest commitments to the past and enables the organization to reexamine its activities with fresh eyes. A wholesale replacement also serves as a signal that this time, at last, the organization is going to make a serious effort to change; after hearing numerous brave speeches that actually led to no major changes, the organization's members have grown cynical and they need a strong signal of sincerity and commitment. Equally important, a wholesale replacement places the blame for failure on the dismissed top managers. This blaming may be unjust, but it lifts the stigma of failure from those who remain in the organization and frees them to seek success in new directions. To turn around, the organization will need their enthusiasm, energetic effort, and belief in their own abilities.

Of course, the organization also needs some new ideas about its capabilities and the opportunities it possesses, but new ideas seem to follow automatically from the new perceptions that come with a change in top management. The organizations we studied did not lack for alternative ideas; many of their members had different viewpoints and had proposed different strategies. Thus, these organizations did not have to import ideas from outside; they looked mainly at old proposals from fresh perspectives. One of the organizations we studied did import some radical new strategic ideas: Its new top managers invited the students in a market research course at a nearby university to study its situation and to propose solutions.

And How Did Facit's Story Turn Out?

You recall that in October 1972, the board of directors discussed McKinsey & Company's proposals for further retrenchment, could not decide what to do, and tried to hire a new managing director who would make the decision, but he refused. At that point, the senior executives believed the company was inexorably headed for bankruptcy. They had no notion what they could do to turn the company around, for they had built a mechanical-calculator company that could not deal with the electronic age. The company was hemorrhaging financially, and they could only visualize more and more factory closings.

At that point, the Electrolux company offered to buy Facit, and the sale went through almost immediately. Later one of Facit's senior executives said that they would have gladly sold out earlier if only someone had made them an offer.

Electrolux took over on a Tuesday. That morning, Electrolux's managing director called all of Facit's senior executives into his office and explained that they would have to depart before Friday. Hearing of this mass firing, a newspaper reporter asked Electrolux's managing director how he was going to replace so many top managers. The managing director explained hyperbolically that Facit represented only a 30 percent increase in Electrolux's sales, and that Electrolux did not think they needed another layer of management to handle a modest increase in sales. Electrolux's action had several implications. From one point of view, Electrolux was putting Facit under the direct supervision of its own top-management team. But Electrolux's senior executives obviously did not intend to drop their previous responsibilities and devote most of their attention to Facit, which was going to operate as an autonomous subsidiary. So from another point of view, Electrolux was passing the top-management responsibility down to the next layer of managers within Facit. Certainly, Electrolux was signaling the birth of a new era.

Electrolux's managing director immediately started making speeches that sounded very different from those heard before. Instead of threats, he spoke confidently of challenges and fast action and the power of a large company. Instead of fears, he talked optimistically about energy and opportunities and the thrill of successful competition.

Electrolux also loaned approximately $2 million to its new subsidiary. For a company of that size, this was a very small sum. The loan mainly symbolized Electrolux's faith in Facit's future.

Looking into Facit's situation, Electrolux discovered that orders for typewriters and office equipment were backlogged many months. The backlogs would probably have been much greater yet, except that salespeople had been turning down orders on the grounds that they could never be filled. The backlogs were so big, of course, because Facit had been closing typewriter and office-equipment factories. So Electrolux inverted this process, closed the mechanical-calculator factories, and converted them to making office equipment and typewriters. While doing this, Electrolux temporarily laid off 800 blue-collar employees, but these people were being rehired within three months. In six months, Electrolux had tripled Facit's typewriter production and doubled its office-equipment production.

Electrolux also discovered that Facit's joint venture with the Sharp Corporation had designed electronic calculators and computers that were ready for production. By July 1973, Facit was making and selling electronic calculators and computer terminals.

Facit broke even on a cash basis in the third quarter of 1973, made a gross profit during 1973, and made a net profit during 1974. During 1974, production rose 25 percent, exports went up 50 percent, and employment increased 10 percent. For a company that had thought it was headed for bankruptcy, this felt really good.

After revisiting the revitalized Facit, a Swedish newspaper reporter wrote: "Although everything looks different today, the company is still more or less managed by the same people who were in charge of the company during the sequence of crises.... All of the present products emanate from the former Facit organization, but still, the situation has changed drastically." But a union officer may have contributed the most insightful observation about Facit's turnaround. He said, "It is again fun to work."

Ten years later, L. M. Ericsson bought Facit from Electrolux to get its expertise as a manufacturer of avant-garde computer equipment.

QUESTIONS AND ANSWERS

Question: You said that of the six case studies of successful turnarounds, none of these were achieved without removing top management. In other instances, have you ever become aware of a successful turnaround of a crisis without replacing top management?

Answer: No, I haven't.

Let me emphasize that it is not just that you must replace some top managers: you have to make a clean sweep across the board. If you try to replace just one or two top managers at a time, you have little impact because the newcomers will learn from their more experienced colleagues. Facit, in fact, did that. They replaced their managing director six times in three years. And every one of the men coming in explained that a new world had opened up, that his predecessor had been lying about how bad things really were, but that he now had the solution and things were going to be different.

Question: You also said that the causes of crises and failures are the same as the causes of successes. Would you take that hypothesis down one step and say that the causes for good management are the same as the causes for bad management?

Answer: I do not know how far it generalizes, but that idea certainly applies to organizations' programs. Good programs may be the roots of your success; but if the world shifts and you keep on performing the same programs, they turn into handicaps.

Question: Would you say as a rule that it is not healthy to promote from within so that all the vice-presidents end up being guys who started on the assembly line and took 20 years to get there?

Answer: I know no percentages, but there are companies that insulate their top-management group with a ring of toadies. Roger Dunbar and Walter Goldberg (1978) argued that such companies are especially prone to get into trouble.

More generally, there is a tendency for promotion processes to filter out dissenters.

Paul Nystrom and I (1984) argue that there are several things that top managers can do to avoid getting into crises in the first place and that it is in the strong self-interest of all top managers—chief executives especially—to do those things. About the worst thing an executive can do is to become a 55-year-old scapegoat for a company going down the tubes.

Among other things, top managers should tolerate dissenters and try to listen down the hierarchy, if they can force themselves to do that. The studies of organizational communications show basically that people talk up and listen up. There is not much communication going downward and not much listening downward. You can see all kinds of reasons why top managers do not want to listen down: They worked hard to get to the top; and now they want to go to the country club and socialize with other people who have gotten to the top. Although this is a natural human thing, I think it is really to the top managers' disadvantage that they do not listen and talk more to their subordinates.

Top managers also ought to run more experiments. They should voluntarily challenge their assumptions by testing them at the periphery. They should constantly ask: "How should we be adapting?" But again, that is a difficult thing to do when they are making a buck, when they have a formula that seems to work. Why are they going to throw away something that works in order to get something that may be worse?

Question: These 12 companies that got into trouble—were they in trouble because of changes in their existing environments? Is that also part of your definition?

Answer: No, environmental change was not one of our criteria. We just looked for signs of trouble, and we looked for different kinds of stories. We found a lot more than 12 companies in trouble; it is just that we only spent time studying 12.

Companies in trouble can always blame their environments, and they can always find somebody inside to blame as well.

Question: I am trying to relate your message to information technology. If I understand correctly, you are saying that companies get into trouble if they do not recognize a change in their environments and they do not change with their environments. Therefore, information technology is one of the major factors, because good information systems help companies to monitor environmental changes.

Answer: One of the characteristics of information technology is that firms design their information gathering and processing systems on the basis of what they believe to be important. Consequently, they build very strong biases into their information systems that tend to preserve the status quo and to blind them to external changes.

Bo Hedberg and Sten Jönsson (1978) wrote an article called "Designing semi-confusing information systems for organizations in changing environments." Their point was that information systems should not remove the uncertainties from information and thus create false senses of security. Information systems should make organizations aware of the limitations in their basic assumptions.

Question: Can we bring this one step further by saying that in some ways, there must be an opportunity for information technology to help companies recognize changes?

Answer: These strategic problems are partly cognitive, and I have thought about getting involved myself in writing some software that would help managers to see their product environments from different perspectives, to see opportunities and alternative strategies. But there are very strong political forces in firms, and merely seeing what to do is not sufficient to produce correct action. People's careers intertwine with strategies, and if you change the strategies, you change power structures of the organizations. The mechanical engineer knows that when the company goes into electronics, it is an electronic engineer who is going to have influence.

REFERENCES

DUNBAR, ROGER L. M., and WALTER H. GOLDBERG (1978). Crisis development and strategic response in European corporations, *Journal of Business Administration* 9(2):139–149.

GRINYER, PETER H., and DAVID NORBURN (1975). Planning for existing markets: perceptions of executives and financial performance, *Journal of the Royal Statistical Society, Series A* 138(1):70–97.

HEDBERG, BO (1981). How organizations learn and unlearn, in Paul C. Nystrom and William H. Starbuck (eds.), *Handbook of Organizational Design*, Oxford University Press, New York, Vol. 1, pp. 3–27.

HEDBERG, BO L. T., and STEN A. JÖNSSON (1978). Designing semi-confusing information systems for organizations in changing environments, *Accounting, Organizations and Society* 3:47–64.

LEACH, EDMUND R. (1964). Anthropological aspects of language: animal categories and verbal abuse, in Eric H. Lenneberg (ed.), *New Directions in the Study of Language*, MIT Press, Cambridge, Mass., pp. 23–63.

MARX, KARL (1904). *A Contribution to the Critique of Political Economy*, Kerr, Chicago.

NORMANN, RICHARD (1971). Organizational innovativeness: product variation and reorientation, *Administrative Science Quarterly* 16:203–215.

NYSTROM, PAUL C., and WILLIAM H. STARBUCK (1984). To avoid organizational crises, unlearn, *Organizational Dynamics* 12(4):53–65.

STARBUCK, WILLIAM H., and PAUL C. NYSTROM (1981). Designing and understanding organizations, in Paul C. Nystrom and William H. Starbuck (eds.), *Handbook of Organizational Design*, Oxford University Press, New York, Vol. 1, pp. ix–xxii.

CHAPTER 2

Identifying Strategic Information Systems: Finding a Process or Building an Organization?

Michael Vitale
Harvard Business School
Blake Ives
Dartmouth College
Cynthia Beath
University of Minnesota

Considerable attention is currently focused on using information technology to obtain and maintain competitive advantage. Numerous mini-cases have been used to illustrate the use of information systems for competitive advantage and various conceptual frameworks have been proposed to aid in the identification of such applications. Much of this work is grounded in a single concept of strategy formulation, an approach that we refer to as "top-down." A survey of senior information system executives demonstrates the potential problems of relying on a top-down approach. A second "adaptive" approach, appears to offer potential value for the identification of competitive applications in organizations facing considerable environmental turbulence or in which senior strategists are relatively uninformed about information system resources. Five organizational roles are defined that can help support this adaptive approach.

Continued decline in the unit cost of computer hardware and steady improvement in the flexibility and power of computer software have led to dramatic growth in the use of information technology in American business. Applications have proliferated rapidly beyond corporate walls to encompass suppliers, distributors, and consumers, and consequently, some applications of information technology have acquired the ultimate label of respectability: they are called "strategic" in business journals (McFarlan, 1984) and featured within, and even on the cover of, widely read business magazines (*Fortune*, 1985; *Business Week*, 1985).

Such attention is welcome legitimization for information systems managers who have worked for years to communicate the exciting potential of information technology. But management's new interest in competitive information systems presents a difficult challenge for information systems managers, who now frequently find themselves charged with responsibility for identifying and implementing strategic applications. Although management attention has been focused on the potential of competitive information systems, there is a dearth of practical, or even theoretical, advice as to *how* to identify such applications. While some information systems have clearly had major strategic impact, at least in the short term, we may be unable, *deliberately and reliably*, to identify and implement information systems that make lasting improvements in competitive position.

Mini-case studies or exemplars, usually in conjunction with a conceptual framework such as Porter's Competitive Forces (Porter, 1980; Porter and Millar, 1985) or Wiseman's Strategic Thrusts (Wiseman, 1985), are currently the method of choice for raising awareness about competitive information systems (McFarlan, 1984; Frohman, 1982; Cash and Konsynski, 1985; Benjamin et al., 1984), facilitating the identification of such systems (Rackoff et al., 1985; Wyman, 1985), and for developing and illustrating new conceptual models (Ives and Learmonth, 1984; Sullivan, 1985). But the endless repetition of a few well-known anecdotes may be an idea generation catalyst of insufficient utility (Bakopoulos and Treacy, 1985), even for the competing airlines or hospital suppliers who found the original stories to be frighteningly motivating. Furthermore, the few most popular anecdotes, like many myths, are often inaccurate and usually exclude details critical to successful imitation. The minicases rarely describe how the initial idea for the system was identified, how the system was initially justified, the costs of the system, or its eventual bottom line impact.

Moreover, much of the literature lends a Pollyanna tone to the issues surrounding strategic information systems; the emphasis has been on describing opportunities and success stories rather than attempting to produce a well-balanced presentation recognizing the high risks often assumed, the false starts undertaken, and the failures that must inevitably occur. Wiseman (1985), for example, identifies 118 competitive information systems, some of which are no longer functioning, but never adequately deals with the potential downside of competitive information systems—the risks assumed, the write-offs incurred, the possibility of competitive retaliation (Cash, 1985).

In what follows we first review some of the existing work on the strategic use of information technology. We are interested primarily in the process of selecting business strategy and the process of connecting business strategy and information technology. We review some data which suggest that the traditional approach may not be appropriate in the IS arena and we then pose an alternative view of this relationship, one that we think will be helpful in thinking about competitively important applications. Finally, we make some suggestions for developing an environment in which strategic information systems will be identified.

THE PROCESS OF STRATEGY DEVELOPMENT

The degree to which information technology is integrated into a firm's strategic planning process varies from firm to firm, with important consequences in industries with information intensive processes. Consider, for example, the airline industry.

The two largest trunk carriers, American and United, are premier examples of what can be done with effectively applied information technology. In both companies information systems reports at a very high level within the organization and plays an integral part in the formulation and implementation of corporate strategy. Both carriers have done well enough with information technology to be sued by competitors on the grounds of unfair competition, alledgedly related to their Sabre and Apollo reservation systems. The cases of Braniff and Frontier contrast markedly with American and United. Immediately after deregulation, Braniff and Frontier both expanded rapidly from their regional niches and began to compete on routes formerly served exclusively by their larger competitors. Neither Braniff nor Frontier had expanded their reservations systems to include travel agents; both were dependent on systems owned by their major competitors for ticket sales and distribution. The relatively infrequent flights that these smaller carriers could sustain to a given city, and the hub-and-spoke arrangements that they, like other airlines, were developing, in fact depended heavily on information technology, but neither carrier seems to have recognized this until it was too late. Finally, at People Express, the granddaddy of the low-cost carriers, the IS function was integrated with the organization in the strongest way possible: the IS staff was made up of pilots, customer service managers, and other line employees. IS applications at People Express include a no-frills reservation system that can be accessed from any touch-tone phone but is not available on travel agents' terminals—an important part of maintaining the airline's low-cost position.

In the airline industry some airlines recognized early that information systems were to play a strategic role. Here and in several other industries, organizations that recognized this strategic role of information systems technology have obtained competitive advantages. We believe that in the near future many organizations will greatly improve their competitive position through the use of information technology—in the long run perhaps to an even greater extent than they can by making more effective use of marketing, logistics, manufacturing, or other functional strengths. The question is: What must we do to ensure that information technology is prudently applied?

The Top-down Model

Much of the existing work on strategic uses of information technology (McFarlan, 1984; Parsons, 1983; Wiseman, 1985) assumes, at least implicitly, a "top-down" model of strategy development (Chaffee, 1985; Chandler, 1962). According to this view,

> strategy consists of integrated decisions, actions or plans that will set and achieve viable organizational goals. Both goals and the means of achieving them are results of strategic decision. To reach these goals, organizations vary their links with the environment by changing their products or markets or by performing other entrepreneurial actions. (Chaffee, 1985, p. 90)

The top-down model assumes that managers are well informed about the organization's resources, in particular, the organization's "distinctive competences" (Selznick, 1957)—the unique strengths of an organization that, if leveraged, can produce competitive advantages that a competitor will be unable to duplicate easily—as well as about the constraints and opportunities in the environment shared with their competitors. This model suggests that strategy selection is a rational response to competitive forces (Porter, 1979) in a relatively predictable world. The next step after strategy selection is to maintain or adjust organizational strengths and weaknesses to the requirements of the new strategy, perhaps, if necessary, by acquiring assets from outside the organization (Andrews, 1971).

In the 1970s, as computing use grew beyond the transaction processing stage, the information systems literature became littered with calls for involvement by information systems managers in the firm's strategic planning process (King, 1978; Kriebel, 1968; McLean and Soden, 1977). The rationale for their participation followed directly from the top-down approach. Involving IS-aware and IS-responsible people during strategy formulation is more likely to lead to applications that bring sustainable competitive advantage because:

- It signals to the organization that top management recognizes the strategic potential of IS.
- It involves IS management at a time when fewer decisions have been made, thus opening up more avenues of exploration.
- It shortens the lead time for new applications, since IS is aware sooner of management's thinking and can start "learning into" the right technologies.

There is nothing surprising here. We would not expect to find successful implementation of a low-cost strategy in a firm that first decided to pursue a low-cost position and then asked how to use marketing and production to support that strategy. In general, however, the appeal for IS involvement fell on deaf ears. IS managers have not been notably successful in interjecting themselves into their firms' strategic planning processes.

As an alternative to direct participation in the organization's strategic planning process, IBM promoted their Business Systems Planning process (IBM, 1984). Using this process, a management systems team infers or recreates the firm's goals and strategies through a series of interviews of structured interviews with top management, for which specific application proposals are eventually derived. The Critical Success Factors (CSF) methodology, another frequently used information systems planning methodology, also circumvents the problem of direct involvement in the firm's strategic planning process. Instead, general managers are asked to identify those "few key areas where 'things must go right' for the business to flourish" for information system designers who are assumed to be ignorant, or at least not well informed about the business (Rockart, 1979, p. 85).

More recently, but still in the spirit of the top-down model, Wiseman (1985) and others (Benson and Parker, 1985) have resurrected the call for IS involvement in strategic planning, suggesting that information technology can "shape" or "impact"—influence the selection of strategy—in addition to "support" or "align"—help the organization realize their strategies.

In the current strategic applications literature, some methodologies for identifying strategic information systems do not work down from explicit statements of goals or strategic thrust. Ives and Learmonth (1984) describe such a method. They suggest that a firm take their customer's perspective in examining the life cycle of activities—requirements, acquisition, stewardship, disposal—that the customer must go through to acquire the firm's product. Frequently, information systems can be used to assist the customer in one or more stages of this "customer resource life cycle," and by doing so, a firm can differentiate its product from that of its competitor.

In a similar vein, Rockart and Scott Morton (1984) focus on information assets, rather than specific goals, in a value-chain analysis of the firm's internal operations and its links to customers and suppliers. In their approach, information technology serves both as the basis for new products and services ("shape") and as a way of linking together value activities ("support"), thereby improving their efficiency and effectiveness.

In general, the typical IS manager does not participate in his or her firm's strategic planning process, but may be under serious pressure from senior management to identify applications that support the firm's competitive strategy. If IS managers are skeptical that the processes proposed to date will, in fact, reveal proposals that will "shape" corporate strategy, it is not surprising. After-the-fact support for corporate strategies developed without regard for information assets or opportunities will rarely result in competitive applications of the technology and usually only by luck. By their very nature these applications are not deeply integrated into the firm's other operations. We identify three reasons that the current top-down-based methods for identifying strategic applications are likely to be unsatisfactory:

1. *Overall strategy does not exist.* Many IS managers would feel very fortunate to have a clear picture of where their organization is headed so that they could match IS and organization efforts. But many organizations have no well-defined strategy. The absence of an explicit strategy will preclude either mapping information systems plans to long-range organizational plans or altering organizational plans through information systems technology.

2. *Strategists are uniformed about information systems technology.* Without sufficient knowledge about information system possibilities and capabilities, organizational strategists have a difficult time making effective decisions about the application of information system technology. Such expertise should include familiarity with the following:

- The current applications portfolio, including knowledge of customer and product data bases, existing information linkages among organizations, cost accounting information, and so on.
- The information architecture, including existing and planned communications networks, distributed processors, planned and existing workstations and personal computers, and so on.
- Systems development management, including a workable understanding of the trade-offs among system costs, schedule and functionality, and a knowledge of the uncertainties connected with system estimates.
- Future technologies or relatively untested technologies that might have a strategic impact on the organization, its products, or its customers

3. *The environment is turbulent.* Considerable change in the information systems environment may reduce the appropriateness of strategies formulated using a top-down process in achieving competitive advantage (Emery and Trist, 1965). Long lead times in the information technology arena increase the risk that a strategy will be valueless before it is fully implemented.

- *Products.* Product changes will become increasingly dependent on information system enhancements as economic considerations drive product designers toward making products more information intensive.
- *Customers.* Customers themselves, and the needs of existing customers will change. Existing customers may insist on electronic linkages for order entry or automated inventory replenishment.
- *Competitors.* Existing competitors may change the basis of competition or distribution channels. Information systems technologies may be used to implement these changes, to introduce barriers to entry to lock out potential competitors, or to break down barriers to entry to potential markets.
- *Suppliers.* With changes in products will come changes in suppliers. New materials technologies, new production methods, and economic pressures will also change the supplier pool. Relationships with existing suppliers will also change. Electronic linkages may prove necessary or competitively advantageous for production planning, inventory control, on-line ordering, quality control, and so on.
- *Production methods.* Changes in production methods may add additional turbulence to the environment. Many of these changes will be related to information system technology including robotics, CAD/CAM, MRP, laser technologies, and so on.

AN EXPLORATORY RESEARCH QUESTION

As an exploratory research question, we propose that in many organizations the environment will not support a top-down process for identifying competitive information systems. To test this notion we surveyed 24 participants drawn from 20 companies attending a conference entitled "Information Processing in American Business." The questionnaire focused on the process that these organizations used to identify strategic uses of information systems technologies.

Responses were received from 17 (71 percent) of the 24 chief information officer or information systems manager participants, representing 13 business units in 11 Fortune 500 (or equivalent) companies, one somewhat smaller company, and three consulting firms. Organizations represented among the respondents included financial services, petroleum, communications, computer vendors (3), business equipment, retailing (2), aerospace (2), food processing, and consulting/public accounting (3). In general, the respondents are highly placed in their organizations. Eleven of 17 hold positions at the vice-presidential level or above, and another four report to vice-presidents.

All but one of the organizations included in the survey have or plan to have a process for identifying strategic uses of information systems technology. Eleven of 17 have a specific ongoing set of activities intended to identify strategic uses of information systems technologies. Of the six who do not currently have a specific process, four plan to implement a specific process within a year, one plans to do so within two years, and one does not plan to do so, having tried it and failed.

Eight of the 11 respondents who currently have a process for identifying strategic applications have used one or more methodologies prescribed for identifying competitive information systems. Seven of the eight had used the Critical Success Factors approach; four of these had also used IBM's Business Systems Planning process; and four had used one of the Porter-based approaches (Competitive Forces, Competitive Strategies, or Value Chain). We also asked the respondents to tell us about specific characteristics of the process currently being used. All 11 include a review of the business strategy, and all but one review the applications portfolio or see presentations of new technologies. Eight review the competitive environment and the company's strength and weaknesses. Based on this information, we argue that the processes being used in these organizations are predominately top down in nature.

RESULTS

Our exploratory research question predicted a relationship between satisfaction with the process for identifying strategic uses of information systems technology and three organizational factors:

1. Knowledge about information assets and opportunities
2. Turbulence in the organization's competitive environment
3. The presence of a strategic plan for the organization

Items addressing these three organizational factors appear at the beginning of the questionnaire. The item measuring satisfaction with the existing system identification process is the last item on the questionnaire. Descriptive statistics for the four measures, for those organizations that have a formal process for identifying strategic applications, are shown in Table 1. In the following sections we review the results for each of the organizational factors.

TABLE 1 Descriptive Statistics

Variable[a]	*N*	Mean	Median	Standard Deviation	Range
Satisfaction	10	3.2	3 (marginally satisfied)	0.60	2–4
Knowledge	10	10.1	10	2.88	6–16
Turbulence	10	14.2	14	3.09	10–20
Unit-level strategic planning: Of 10, 9 responded "yes," 1 responded "no"					

[a]The knowledge and turbulence scores were obtained by assigning values to subitem responses and summing these values. Values were assigned to the responses as follows:
Knowledge:
1 = uninformed
2 = somewhat informed
3 = well informed
4 = very well informed
(possible scores range from 4 to 16)
Turbulence:
1 = no change
2 = minimal change
3 = moderate change
4 = major change
(possible scores range from 4 to 20)

Presence of a Strategic Plan

Considering the size and maturity of the organizations represented, it is not surprising that we found the level of formal planning to be relatively high. All but two of 17 do strategic planning, and all but two do information systems planning. Given the lack of variance in the responses to this item, we did not include it in our further analysis, but the prevalence of strategic plans does lend some support to the potential utility of the top-down model for identifying strategic applications for most of the organizations included in the survey.

Knowledge about Information Technology

We also predicted that the top-down approach will be most appropriate if the organization's strategists are knowledgeable about information assets and information opportunities. The underlying assumption here is that managers with limited understanding of information system possibilities and constraints will be unable to formulate feasible and effective information-based strategies. As shown in Table 2, the data convincingly demonstrate that there is a strong relationship between IS knowledgeable management and system identification processes that are viewed as satisfactory (a Pearson product-moment correlation coefficient of 0.86 is statistically significant at $p < 0.001$).

TABLE 2 Pearson product-moment correlation coefficients

	Satisfaction r	r^2	t	p
Knowledge	0.86	0.73	4.70	<0.001
Turbulence	–0.61	0.38	–2.20	<0.05

Environmental Turbulence

Our third prediction was that environmental turbulence associated with frequent changes in product, supplier customer, production processes, or competitive environment would make top-down planning more difficult, if not irrelevant, and should be associated with lower levels of satisfaction with the planning process. As shown in Table 2, this prediction was also validated by the survey data. Organizations facing high levels of environmental turbulence were less satisfied with their existing process for identifying strategic applications of information systems.

Summary

In an environment characterized by considerable turbulence vis-à-vis changes in products, customers, channels, and underlying production technologies, a top-down planning process is susceptible to wasted efforts, misdirected investments, and low morale as the organization pursues strategies based on incomplete or inaccurate predictions of the future. Clearly, our data suggest that, at least in the organizations surveyed, the respondents perceived high levels of environmental turbulence. Moreover, in those organizations experiencing the most turbulence, there was a much higher degree of dissatisfaction with the process being employed to identify competitive applications.

High levels of environmental turbulence cast shadows on the utility of the top-down planning process as an instrument for identifying competitive applications of information system technology. So, too, do senior strategists with limited understanding of information technologies and resources. Information technology can "shape" the firm's strategy only if consideration of information assets and opportunities is incorporated into a firm's strategic planning process in some way. Technological opportunities available in the marketplace to all competitors in an industry (personal computers, commercial software, optical cables) may permit short-term strategic gains or catch-up moves. But to achieve defendable, long-term strategic benefit, those external opportunities will probably have to be linked to unique assets currently among the organization's resources, and, most likely, to the organization's distinctive competences, or the factors that give it an advantage over its competitors. That is, pure technology probably offers little strategic advantage; advantage flows from using technology to leverage or exploit inimitable assets of the firm—its history, reputation, culture, management skills, market power (Barney, 1983). That is why, as our data suggest, it is so important for strategists to be familiar with information technology opportunities.

In addition, the firm must recognize that it may have exploitable information assets—historical customer or supplier data, order-entry and production application linkages, communication infrastructure, system development teamwork skills—which also might be used as the basis for sustained strategic advantage, either in combination with new information technology opportunities, or in combination with other distinctive competences or unique assets of the firm. Again, this suggests the need for strategists to be familiar with their current information assets.

THE ADAPTIVE MODEL

A second model of strategy, not as yet explicitly referenced in the strategic IS applications literature, emphasizes continuous adaptation, creating "satisfactory alignments of environmental opportunities and risks, on the one hand, and organizational capabilities and resources, on the other" (Miles and Cameron, 1982, p. 14). This "adaptive" model (Chaffee, 1985), unlike the top-down model, deals less with predetermined goals, and more with evolution and incremental change.

Moreover, "the adaptive model's definition of strategic behaviors goes beyond that of the top down model to incorporate not only major changes in products and markets, but also subtle changes in style, marketing, quality, and other nuances" (Chaffee, 1985). Quinn (1980) contends that the adaptive model seems closer to the way most firms actually produce and implement strategy than the top-down model.

Hayes (1985) suggests that managers typically approach the problem of strategic planning the wrong way around; it makes more sense, he says, to develop resources, and then identify tactics and strategies that are feasible. This concept is highly relevant to the information systems environment. For example, he suggests that the firm "acquire and experiment with new technologies and techniques so that workers and managers gain experience with them and come to understand their capabilities and constraints" (p. 118). The broadly based knowledge that results contrasts sharply with the pockets of expertise that are typical with the top-down approach.

Hayes suggests mechanisms for dealing with the turbulence problem. As knowledge spreads throughout the organization, "the company should encourage managers well down in the organization to exploit matches wherever they occur" (p. 118). The point is to "seek continuous improvement in a dynamic environment" by reacting quickly to perturbations in the competitive environment at relatively low levels of the organization.

The top-down model for identifying competitive information systems is normally implemented as a *process.* Hayes advice suggests, instead, changes in organization *structure* to facilitate the development of technical means and the innovative application of those means in exciting new ways. For example, rather than relying on a specific process for identifying competitive applications, with its attendant roles of "process facilitator," "executive sponsor," and "participant," Hayes approach suggests that permanent changes be made to the organization. Some alternative permanent roles seem to offer potential for helping to integrate information technology and strategic planning. Illustrative role descriptions, described below, draw from the works of Maidique (1980) and Morrison (1950) on managing technical innovation.

Wizard. Wizards are the corporate experts and librarians for technologies that are judged to have potential strategic impact on the firm. Such people are charged with familiarizing themselves with how the technology operates, what its costs and capabilities are today, what its costs and capabilities are expected to be next year or in five years, what the limitations of the technology are, and how it compares with substitute technologies in terms of costs, capabilities, and limitations. Wizards should also be aware of the organization's own strengths and weaknesses vis-à-vis the technology. They should also maintain (or contribute to) a library of expertise about the technology, including materials suitable for potential management readers. Finally, experts should maintain files of example applications of the technology, including applications within or external to the organization. Periodically, wizards should prepare short summary papers on the technology consistent in format with papers prepared for other technologies and understandable to a general management reader.

Marriage Broker. One or more relatively senior information systems executives might be designated to act as intermediaries between user managers with interesting ideas and the wizards. These marriage brokers will be charged with the responsibility for seeking out and encouraging new ideas for competitive applications, ensuring their technical feasibility, and nurturing budding systems. Like the original marriage broker, ours carefully avoids meetings between the user and the wizard before the plan has reached a certain level of maturity. Our marriage broker serves not as a dating service but is instead a translator, chaperone, and even, where necessary, a barricade. The marriage broker should ensure that library materials made available to a prospective user are suitable for his or her needs and background and that the applications presented to the user are understandable and appropriate.

Rich Uncle. Money needs to be available to acquire experimental technologies that appear to have some potential merit. Prototype applications need to be supported that might not otherwise pass the organization's investment hurdles. The rich uncle pays for seeds that users plant, in hopes that trees will grow. These "research and development" funds need to be available in proportion to the perceived strategic importance of information systems to the organization.

Weed Puller. Sometimes our seeds produce weeds instead of trees. But prototypes and experimental technologies may take on fascinating lives of their own that far exceed their utility to the organization. Periodically, such investments must be reexamined and their current utility assessed. The weed puller must be in a position high enough in the organization to be able to avoid the pressures of the status quo or of special constituencies. He or she also has a responsibility to ensure that those who grew the weeds are not branded as having somehow failed, but rather are viewed as temporarily unlucky horticulturists.

Teacher. The teacher plays an important role in the adaptive organization. Systems analysts, wizards, and others within the information systems group must learn more about the organization's products, customers, competitors, suppliers, and distribution channels. The teacher provides an environment for this learning. The teacher also educates users about the possibilities presented by various technologies. Although the subject of such lessons is ostensibly technology, the presentations cannot be technical. They need to be short introductions to the technology, with most emphasis placed on existing applications and the identification of potential applications.

Wizards are typically no better suited to be teachers than they are to staff the information center. Moreover, teachers need high levels of credibility with their intended audience. In one organization we are familiar with, the head of information systems sees his most important role as being the technology teacher to the management committee. He invests considerable time (and patience) into ensuring that senior executives are aware of the potential of cutting-edge technologies. Obviously, external experts and consultants frequently fill this role. Job rotation policies, rigorously adhered to, may provide some of the function of the teacher.

CONCLUSION

Both the literature on strategic information systems and current management practice are strongly focused on developing mechanisms for identifying strategic applications. The idea generation processes in use are all illustrative of the top-down approach—senior strategists attempting to fit information systems technology to organizational goals or using technological possibilities to derive organizational goals. Unfortunately, our data suggest that such strategies are not likely to be particularly satisfactory if the senior strategists are relatively unaware of technical possibilities, or if the organizational environment is particularly turbulent. The current focus on process issues risks ignoring the distinctive information-related competences of the firm, thus potentially overlooking opportunities that might result in sustainable competitive advantage. Development of an organizational structure conducive to the identification and implementation of strategic applications may, in the long run, prove to be a sustainable competitive advantage in its own right—that is, an organizational structure that continually assures that the firm is the leader in the strategic use of information technology.

REFERENCES

ANDREWS, K. R. (1971). *The Concept of Corporate Strategy*, Richard D. Irwin, Homewood, Ill.

BAKOPOULOS, J. A. Y., and M. E. TREACY (1985). Information technology and corporate strategy: a research perspective, *CISR Working Paper 124*, Center for Information Systems Research, MIT, Cambridge, Mass.

BARNEY, JAY B. (1983). *Strategy and Organization*, UCLA Graduate School of Management, Los Angeles.

BENJAMIN, R. I., J. F. ROCKART, M. S. SCOTT MORTON, and J. WYMAN (1984). Information technology: a strategic opportunity, *Sloan Management Review* 25(3):3–10.

BENSON, R. J., and M. M. PARKER (1985). Enterprise-wide information management: an introduction to concepts, *Enterprise-Wide Information Management Workshop Report,* June 19–21, Washington University, St. Louis, Mo.

BUSINESS WEEK (1985). Information power, October 14, pp. 108–116.

CASH, J. I. (1985). Interorganizational systems: an information society opportunity or threat? *The Information Society* 3(3):199.

CASH, J. I., and B. KONSYNSKI (1985). IS redraws competitive boundaries, *Harvard Business Review* 63(2):134–142.

CHAFFEE, ELLEN E. (1985). Three models of strategy, *Academy of Management Review* 10(1):89–98.

CHANDLER, A. D., JR. (1962). *Strategy and Structure,* MIT Press, Cambridge, Mass.

EMERY, F. E., and E. L. TRIST (1965). The causal texture of organizational environments, *Human Relations* 18(February):21–32.

FORTUNE (1985). How to keep your customers happy captives, September 2, pp. 42–46.

FROHMAN, A. L. (1982). Technology as a competitive weapon, *Harvard Business Review* (January–February):97–104.

HAYES, ROBERT H. (1985). Strategic planning—forward in reverse? *Harvard Business Review* 63(6):111–119.

IBM (1984). *Information Systems Planning Guide: Business Systems Planning,* 4th ed., IBM Corporation, New York.

IVES, B., and G. LEARMONTH (1984). The information system as a competitive weapon, *Communications of the ACM* 27(12):1193–1201.

KING, W. R. (1978). Strategic planning for MIS, *MIS Quarterly* 2(1):27–37.

KRIEBEL, C. H. (1968). The strategic dimensions of computer systems planning, *Long-Range Planning* (September).

MAIDIQUE, MODESTO A. (1980). Entrepreneurs, champions, and technological innovation, *Sloan Management Review* 21(2):59–73.

MCLEAN, E. R., and J. V. SODEN (1977). *Strategic Planning for MIS,* Wiley, New York.

MCFARLAN, F. W. (1984). Information technology changes the way you compete, *Harvard Business Review* 62(3):98–103.

MILES, R. H., and K. CAMERON (1982). *Coffin Nails and Corporate Strategies,* Prentice-Hall, Englewood Cliffs, N.J.

MORRISON, ELTING E. (1950). A Case Study in Innovation, *Engineering and Science Monthly* (April).

PARSONS, GREGORY L. (1983). Information technology: a new competitive weapon, *Sloan Management Review* 25(1):3–14.

PORTER, MICHAEL E. (1979). How competitive forces shape strategy, *Harvard Business Review* 57(2):137–145.

PORTER, M. E. (1980). *Competitive Strategy: Techniques for Analyzing Industries and Competitors,* Free Press, New York.

PORTER, MICHAEL E., and VICTOR E. MILLAR (1985). How information gives you competitive advantage, *Harvard Business Review* 63(4):149–160.

QUINN, J. B. (1980). *Strategies for Change: Logical Incrementalism,* Richard D. Irwin, Homewood, Ill.

RACKOFF, N., C. WISEMAN, and W. A. ULLRICH (1985). Information systems for competitive advantage: implementation of a planning process, *MIS Quarterly* 9(4):285–294.

ROCKART, J. (1979). Chief executives define their own data needs, *Harvard Business Review* 57(2):81–92.

ROCKART, J. F., and M. S. SCOTT MORTON (1984). Implications of changes in information technology for corporate strategy, *Interfaces* 14(1):84–95.

SELZNICK, P. (1957). *Leadership in Administration: A Sociological Interpretation,* Harper & Row, New York.

WISEMAN, C. (1985). *Strategy and Computers: Information Systems as Competitive Weapons,* Dow Jones–Irwin, Homewood, Ill.

WYMAN, JOHN (1985). Technological myopia—the need to think strategically about technology, *Sloan Management Review* 26(4):59–64.

CHAPTER 3

Seeking Competitive Advantage Using Information-Intensive Strategies: Facilitators and Inhibitors

William R. King
Varun Grover
Ellen Hufnagel
University of Pittsburgh

In 1981, the senior author wrote a paper entitled "Information as a strategic resource" (King, 1981) that described how firms such as American Airlines and Merrill Lynch were using information technology to gain a comparative advantage in the marketplace. The paper was reviewed by the *Harvard Business Review* and rejected.

Apparently, this was an idea whose time had not quite come in the opinion of *HBR*. However, soon after, they and others began publishing papers that described how these firms and others had successfully used information technology to influence their competitive situation. Subsequently, this has become one of the most discussed ideas in the information systems area in the 1980s. Acceptance of the notion that computers have finally "moved out of the back room" and become an important organizational resource, comparable to the organization's human and capital resources, is potentially the most significant occurrence in the computer era.

Only a few years ago, computers and information were largely seen as service activities that facilitated, but did not importantly influence, the operations of most firms. With the exception of firms that are in the "information business" and those in which computers represent the "core technology" of the business (e.g., commercial data-base services and banks respectively), most firms treated computer operations as a necessary expense item. The computer budget was typically controlled as were other expense categories (e.g., heat, electricity, etc.) in terms of year-to-year changes and in terms of the percentage relationship to sales revenues (or some other such ratio). Investments in computer systems were typically justified on the basis of projected cost savings, with little attention being explicitly given to the potential for added revenues or other benefits that might accrue from them.

Concurrent with the proliferation of microprocessors in the form of personal computers and word processors throughout the organization came the realizations that investments in information resources were growing rapidly, and that there was great potential for these investments to produce substantial benefits rather than merely to enhance efficiency. Anecdotes involving firms that had successfully achieved such benefits began to proliferate, and the notion of information and information technology as a "strategic resource" or source of potential "competitive advantage" became commonplace (King, 1981, 1983, 1984).

Unfortunately, although this notion is potentially important and powerful, it is also imprecise and unstructured. Most of the literature in the area is anecdotal and the remainder is primarily prescriptive. There is little evidence of an accepted or evaluated theoretical framework or structure for the idea (although a wide variety of frameworks have been suggested to be adaptable to this notion) and even less in the way of nonanecdotal empirical evidence.

In this paper an attempt is made to consider some of the fundamental issues surrounding the notion of information as a strategic resource as well as to report on some empirical evidence as to what is actually being done by firms in making these concepts operational. In particular, we seek to:

- Distinguish between "strategic" uses of information resources and uses for purposes of "competitive advantage"*
- Make the distinction between the use of information and of information technology in these fashions
- Report on the *specific ways* in which information resources are actually being used for these purposes by business firms
- Report on the planning and decision-making *processes* that are being used to identify and select opportunities for these uses of information resources
- Report on organizational factors that *facilitate or inhibit* these uses of information resources

*We shall use the term "strategic uses" to be all-encompassing for the moment. Subsequently, we shall distinguish between strategic uses and uses directed toward competitive advantage.

INFORMATION RESOURCES: INFORMATION VERSUS INFORMATION TECHNOLOGY

One of the most obvious things that can be inferred from the anecdotes that appear in the literature concerning strategic uses of information resources and the use of these resources to achieve competitive advantage is that information and information technology are often discussed interchangeably. Since "information technology" can be thought of as a resource that is fairly well defined and readily measurable—in terms of the number, memory capacity, transmission capacity, processing capacity, and so on, of computer-based systems—and "information" is a quite different resource that is less well understood and much more difficult to measure (King and Epstein, 1976), it appears to be necessary to make this distinction as clearly and explicitly as is possible if a useful conceptual framework is to be developed. This is particularly important since the potential applications may be quite different for the strategic uses of information technology than for the strategic uses of information.

There is therefore the need to distinguish between two varieties of *information resources*—information technology and information. *Information technology* can be defined as the hardware and software that is used to collect, transmit, process, and disseminate data (symbols) in an organization. *Information,* according to a traditional (nonoperational) definition, is "data that have been evaluated for some use." In most organizations, there are a variety of "evaluation" filters that are applied to data. Thus, there are a variety of levels of information: for example, those data that have been evaluated for and thereby incorporated into, a data base, those data that are further evaluated and are thereby routinely reported to division managers, and so on.

Clearly, the ways in which information technology can be used as strategic resources may be quite different from the ways in which information can be so used. Possibly, the processes that may be effectively employed to identify such opportunities and the organizational factors that facilitate or inhibit these processes are also different.

Some companies have used information technology to achieve strategic objectives. Taking advantage of technological advances in computing, they provide new ways of relating to buyers or to suppliers to gain advantage over competitors. For example, firm A determined that relations with its customers could be strengthened by placing terminals in each customer location to allow for easier order placement as well as providing a means of advance price and availability checking. While this information was previously made available to customers by phone on request, it is now provided instantly via communication links. Customers are thereby "tied into" firm A in a fashion that makes it more difficult for them to change suppliers.

In the case of firm A, it is the *information technology* and *not the information itself* (since the same information was previously available), that was the key factor in achieving improved customer relations and/or competitive advantage.

Other companies are *using information itself,* rather than information technology, as a key resource to support overall business strategy and to strengthen their competitive position. For instance, firm B is employing a strategy of expansion through acquisition of related businesses. Having identified criteria for evaluating potential acquisitions, the company now scans a commercial data base to identify acquisition prospects that meet the criteria. In doing so, they are *using new data or using available data in new ways* to widen the range of firms that are considered and to make better strategic acquisition choices.

In firm B, it is the *use of important new information or of available information in new ways* that is the key to success. Although computer technology facilitates the processing of this information, it is not the technology itself, as in the case of firm A, that provides the advantage.

The primary distinction between these two approaches lies in the source of the value added. In the first example, the *introduction of technology* increased the value of existing information by providing easier access or speeding up transfer time. In the second, the *information itself* provides a strategic advantage through the new uses to which it is put.

Of course, in some situations, *both* information technology and information can be used to gain competitive advantage. Airline C, which has already pursued the use of *information technology* by putting computer reservation terminals into travel agencies, now begins to use the detailed *information* that it has thereby obtained on supply and demand for various routes to manage optimally the availability of low-price seats on each of its flights. This ensures that it will capture much of the business traveler's demand for high-priced unrestricted tickets as well as fill otherwise-empty seats with low-priced tickets that are sold with restrictions of timing, penalties for changing reservations, and so on. Thus, airline C has used *both information technology* (the terminals in the agents' offices and the associated software for making reservations) and *information* (the supply–demand data for various routes and the associated analytic software) as ways of gaining strategic advantage.

It may be argued that the strategic uses of information technology and information are interdependent. For instance, in the airline example, the strategic use of the technology was a necessary precursor to the strategic use of the information. However, if the same information had been available through some other means than the owned reservation system, the same use could have been made of it. This situation actually exists since some airline reservation systems generate additional revenues by selling such information to other airlines.* For the purchaser of this information, it is information, and not information technology, that may be used in ways that make it a strategic resource.

The important point is that in our failure heretofore to draw careful distinctions between strategic uses of information technology and such uses of information, we have stifled efforts to develop frameworks and processes for guiding our thinking. This is so because information technology is inherently different from information. It has a physical being; it is tangible; it is easy to assess and measure by counts, memory capacities, processing speeds, and so on. Further, it obeys certain rules that information does not.

Information, on the other hand, is intangible and hard to assess or measure. The literature on the "value of information," for instance, is vast, but no single agreed-on measure exists of information value (King and Epstein, 1976). Moreover, information does not obey some important "conservation" rules. If one uses a financial resource, it is no longer available for alternative uses; if one uses information, it is still available. If one uses a portion of a physical resource, the remainder invariably has less value than the original amount. The use of information may even add value to the information that "remains."

Since information technology and information are so different, it may be that the two entities should be considered and dealt with separately if the best results are to be achieved. When managers in a firm attempt to identify strategic uses of information resources, for instance, ideas that relate primarily to information technology will usually emerge first and most prolifically because it is easier to visualize and deal with these tangibles. The intangible information resource will probably be pushed into the background, when in fact, it may have the greater potential to produce benefits. So it is important to make the distinction, if only because it helps us to think in different terms, with different models and expectations, about the two varieties of information resources.

There is one problem with this neat dichotomy, however—the role of software. Clearly, some software adds value in very much the same way as does hardware (e.g., operating systems, compilers, and computer languages). These kinds of software can probably be assessed very much in the same terms as we use to value hardware: processing speed, reliability, and so on.

Other software (e.g., applications software) needs to be assessed primarily in terms of the *use* to which it is put, the task that it performs, and the benefit that accrues from it in the task context—very much like information.

In the case of the earlier-mentioned airline using both information and information technology strategically, the reservation software is very much part of the information technology. The software that is used to analyze the supply–demand data has attributes that are much like that of information.

---

*This raises the interesting management questions of how to price such information, whether to "cripple" it, and so on.

So, the neat dividing line between information technology and information is not completely clear, particularly at a time when even the distinction between hardware and software is itself becoming nebulous. Nonetheless, we believe that the information technology-information dichotomy is useful to guide one's thinking about strategic opportunities.

THE STRATEGIC USE OF INFORMATION RESOURCES

Whichever variety of information resource is being dealt with, the literature has not clearly distinguished between *strategic* uses of information resources and those *uses that are aimed at using information resources as "competitive weapons." A strategic use is one that is aimed at achieving a sustainable comparative advantage; a "competitive weapon" is anything that is useful in competing in the marketplace.*

Because the literature has been so imprecise, it is not surprising that managers tend to discuss specific opportunities or accomplishments in the "strategic" domain in terms that do not really apply. How, for example, can a new inventory control system be used to achieve comparative advantage? While many such "organizational support" systems are discussed in the literature, they are generally aimed at achieving greater organizational efficiency. They are not generally strategic, in the sense that they are not a key element of the organization's strategy and they do not notably influence its ability to achieve a *sustained comparative advantage*—which is the objective of all strategy.

That is not to say that an inventory control system *cannot* be strategic or important to the achievement of comparative advantage; it is merely to say that most such descriptions in the literature are justified not on these grounds but on grounds of being complex and expensive applications of information technology. Implicitly, being so, they "must" be strategic! However, it just isn't so! For an inventory control system to play this role, it must play a direct role in the implementation of the business strategy and the achievement of comparative advantage. Such might be the case if product availability was a critical success factor for the business and/or if inventory costs were a major cost element for a firm seeking to compete as a low-cost producer.

In most other cases, an inventory control system, however complex, sophisticated, and costly it may be, is neither strategic in terms of importance to the organization's goals nor more than a secondary element in its seeking of comparative advantage. Such systems perform traditional support roles, much as the heat and light systems perform support roles. They may be "competitive weapons" in the sense that they help in the competitive struggle, or even because they can be "pointed to with pride" to impress customers or to help close sales, but they are seldom really strategic in nature.

In this manner of equating strategic with "important" or "complex" or "expensive," banks refer to their automatic teller systems (ATMs) as strategic even in cases where they are being installed belatedly merely to "keep up" with competitors who have already installed them. Rather than the creation of comparative advantage, such applications, at best, have the purpose of not surrendering such advantage.

To be a strategic use of information resources, the information resource application must be a significant element of the business strategy—the way in which the firm seeks to gain a sustainable comparative advantage in the marketplace. Although the term is defined only imprecisely, the anecdotes in the literature suggest that a system can be a "competitive weapon" on the basis of much less potential impact.

THE EMPIRICAL STUDY

Having conceptualized the distinction between information technology and information and identified "truly strategic" applications of information resources, we sent out to determine the ways in which these applications were actually being carried out by business firms. As well, we wished to determine the business processes within which these decisions were being made and any organizational characteristics that might serve to facilitate or inhibit these processes.

The data were gathered using a mail survey of information executives (members of the Society for Information Management). The questionnaire contained items that were aimed at assessing:

- The relative pervasiveness of the strategic uses of information resources in various contexts
- The organizational processes that are used to identify and evaluate the opportunities for strategic uses of information resources
- Organizational characteristics that facilitate or inhibit these processes

The specific measures used for each of these categories are discussed below.

Identifying Strategic Uses of Information Resources

A variety of frameworks have been prescribed to aid managers in identifying potential strategic uses of information resources. The most fundamental of these appears to be the well-known model of Porter (1980), which describes the competitive business environment in terms of primary "forces."

We determined that, because of the great face validity and the widespread awareness of Porter's forces, an adaptation of this notion would be used as a basis for data collection concerning the specific strategic applications that firms had made of information and information technology. Thus, we asked respondents whether or not their firm had made strategic uses of either information technology or information in the areas of:

- Supplier relations
- Customer service
- Product/service differentiation
- New product planning
- Cost competitiveness
- Market segmentation

In each such question, we gave an example of what such an application might be expected to do and we asked respondents to provide their own specific examples to substantiate any positive responses that they gave.

Because of the potential importance of the use of information technology in the important area of environmental scanning (Fahey and King, 1977), we asked similar questions in that domain. We also asked respondents to identify the *relative importance* of these areas, both currently and in the future, by selecting the top three most important areas.

Processes for Developing Strategic Uses of Information Resources

We are also interested in the organizational processes and systems that might be used to address these important issues. Therefore, we asked whether the value-added chain (Grant and King, 1982) was used as a basis for identifying such opportunities. We also asked about the frequency and nature of the process used now and three years ago by each firm and whether any formal processes existed for:

- Soliciting ideas
- Funding the exploration of ideas
- Providing administrative support
- Measuring the benefits that are expected to result, or which do result, from uses of information or information technology to gain comparative advantage

Facilitors and Inhibitors

From the literature of implementation, we identified organizational factors that have been studied as facilitators and inhibitors in various contexts. We reviewed these for overlap and redundancy and adjusted them to apply to the information and information technology environment. The list of inhibitors is shown in Table 6 and the list of facilitators in Table 7.

Respondents were asked to judge the degree to which each was an inhibitor or facilitator on a five-point scale ranging from "greatly inhibiting" to "not an inhibiting factor" and "greatly facilating" to "not a facilitating factor."

RESULTS OF THE STUDY

Table 1 shows the industry distribution of the first 51 responses that have been received. Clearly, manufacturers are represented more heavily in these early responses than in the sample to which the survey was mailed, indicating a possible bias in the early responses. That should be kept in mind when interpreting these preliminary results.

Table 1 Respondents by Line of Business ($n = 51$)

Manufacturer	45.10
Finance, insurance, real estate	11.76
Medicine, law, education	1.96
Wholesale or retail trade	7.84
Bus services	3.92
Government	1.96
Public utility, communication, transportation	11.76
Mining, construction, Pct/agriculture	13.73
Publisher	1.96

Areas of Opportunity

Table 2 shows the summary of yes–no responses of 51 respondents to the six areas of potential strategic application for *information technology.* Over 70 percent of respondents claimed to have made applications in the area of *customer service* and nearly 65 percent claimed applications in *cost competitiveness.* Surprisingly, only slightly more than 31 percent claimed applications in supplier relations.

Table 2 Areas of Application for Information Technology as a Strategic Resource ($n = 51$)

Area	No	Yes	No Response
Customer service	27.45	70.59	1.96
Supplier relations	68.63	31.37	—
Product/services differentiation	54.90	45.10	—
New product planning	60.78	35.29	3.92
Market segmentation	60.78	39.22	—
Cost competitiveness	33.33	64.71	1.96
Other	66.67	29.41	3.92

Table 3 shows the same summary with respect to strategic *information-based* applications. While the largest proportion is also in the area of customer relations, the profile is quite different. Only 37.25 percent claim applications of information to cost competitiveness (compared with 65 percent for applications of information technology) and almost 53 percent had applied information to market segmentation.

Table 3 Areas of Application for Information as a Strategic Resource ($n = 51$)

Area	No	Yes	No Response
Customer service	27.45	68.63	3.92
Supplier relations	54.90	45.10	—
Product/services differentiation	60.78	39.22	—
New product planning	60.78	37.25	1.96
Market segmentation	45.10	52.94	1.96
Cost competitiveness	62.75	37.25	—
Other	78.43	13.73	7.84

So, while the primary application of both information technology and information is to customer service, the second-most-prominent area of such applications are quite different.

Organizational Processes

With regard to the process used in the organization to identify opportunities for the application of information technology and/or information for competitive advantage, only 13.73 percent of respondents indicated that they used the value-added chain framework.

When asked to describe the frequency of the process that they used as either continuous, regular, or irregular (Fahey and King, 1977) the responses were those shown in Table 4. Nearly 59 percent responded either "irregular" or "no formal process," suggesting that *analyses dealing with information technology and information as a strategic resource have not yet become a routine and regular part of the planning process.* This is confirmed by the responses to a question of whether these analyses were done as a regular element of the IS planning process, the business planning process, or "no formal" process, where 31.37 percent indicated "IS planning," 13.73 percent indicated "business planning," and 52.94 percent indicated "no formal process."*

Table 4 Frequency of Process for Dealing with Strategic Information ($n = 51$)

Continuous	Regular	Irregular	No formal
21.57%	19.61%	37.25%	21.57%

As to whether formal approaches exist for idea generation, funding, administration, and result measurement, only administrative support stands out at 37.25 percent, compared to less than 30 percent for the other areas.

These overall results suggest that formal processes have not yet been widely developed for identifying and supporting the development of ideas for using information technology and information as a strategic resource.

Relative Importance of the Opportunity Areas

Table 5 shows the areas of opportunity as they were ranked in terms of *relative importance* by the respondents. As expected, 56.86 percent ranked customer service first and only 7.84 percent omitted it from their top three. Surprisingly, supplier relations and new product planning, two of the areas commonly suggested as particularly fertile ones, were omitted from the top three by more than 80 percent. Also surprisingly, market segmentation, which was second in terms of frequency of application for information, was not ranked in the top three by nearly 40 percent of respondents. This suggests that this area may be perceived as having already been "milked" for its primary benefits.

*Note that the "no formal" option was provided as a rough consistency check in two different questions dealing with the "frequency" and "nature" of the process. Given the imperfect distinction between the "irregular" and "no formal" responses to the frequency item, the responses seem to be quite consistent.

Table 5 Relative Importance of Areas of Application

Area	First	Second	Third	Other
Customer service	56.86	25.49	7.84	7.84
Supplier relations	0.00	11.76	5.88	80.39
Product/services differentiation	13.73	29.41	17.65	37.25
New product planning	3.92	5.88	1.96	86.27
Market segmentation	3.92	9.80	13.73	70.59
Cost competitiveness	13.73	17.65	27.45	39.22

Inhibitors

Thirteen possible inhibitors that may have "actually prevented or inhibited your company's efforts to utilize information or information technology for strategic purpose" were identified and respondents were asked to rate each on a five-point scale. Table 6 shows the results.

The relative importance of this area as a strategic priority for the organization, the lack of appropriate planning, the lack of top management support, and the difficulty in assessing the tangible contribution stand out as major inhibitors. These strongly suggest that information resource-based areas of opportunity still have not been focused on by top management,

Table 6 Organizational Inhibitors

Inhibitor	Mean[a] (Standard Deviation)
Lack of appropriate planning	3.412 (1.329)
Low perceived importance of concept	3.039 (1.296)
Lack of appropriate technical support staff	2.294 (1.188)
Budgetary constraints	3.157 (1.155)
Difficulty in assessing tangible contribution	3.451 (1.045)
Complexity of the concept	2.922 (1.111)
High potential startup difficulties	2.980 (1.01)
Lack of organizational/top managment support	3.333 (1.409)
Power and politics in the firm	3.078 (1.426)
Nature of external environment or industry	2.608 (1.250)
Ill-defined management objectives	3.059 (1.173)
Other priorities (are more important)	3.608 (1.115)
Other (please specify)	4.75 (0.50)

[a]0, Not inhibiting; 5, greatly inhibits.

through priorities that have been set through integration into the planning process, and through the development of measurement paradigms.

Thus, while the area is much discussed, it is obvious that much remains to be done in integrating information-based considerations into the strategic milieu of the organization.

Facilitators

With respect to facilitators, the 10 items in Table 7 were identified. The primary facilitators were, as expected, strong technical support and expertise, the leadership position in information technology, competitive pressure, financial position, and computer facilities. Thus the existence of appropriate resources and competitive pressures appears to be the strongest facilitator to effective performance in this area.

Table 7 Organizational Facilitators

Facilitator	Mean[a] (Standard Deviation)
Strong market position of the firm	2.863 (1.386)
Existing information technology leadership position	3.157 (1.239)
Strong planning capability of the firm	2.412 (1.080)
Extensive computer facilities within the firm	3.137 (1.200)
Strong organizational/top management	2.529 (1.317)
Pressure from competition	3.157 (1.206)
Strong technical support/expertise within the firm	3.360 (1.064)
Strong financial position of the firm	3.059 (1.302)
Need for uniqueness or innovation	2.686 (1.140)
Other (please specify)	4.250 (0.957)

[a]0, Not facilitating; 5, greatly facilitates.

SUMMARY

The distinction between "information" and "information technology" and that between "strategic" applications and those that merely constitute the use of information resources as "competitive weapons" are posited as being important to the further development of rigorous procedures and research in the area of strategic information resources.

Preliminary analysis of an empirical survey suggests that:

- The areas in which information technology and information have found strategic application are, in fact, different.
- Most organizations have not yet instituted formal processes for assessing and implementing business strategies that are based on information resources!

REFERENCES

FAHEY, L., and W. R. KING (1977). Environmental scanning in corporate planning, *Business Horizons* 20(4):61–71.

GRANT, J., and W. R. KING (1982). *The Logic of Strategic Planning,* Little, Brown, Boston.

KING, W. R. (1981). Information as a strategic resource, *Working Paper 504,* Graduate School of Business, University of Pittsburgh, Pittsburgh, Pa.

KING, W. R. (1983). Information as a strategic resource, Editor's Comment, *Management Information Systems Quarterly* 7(3).

KING, W. R. (1984). Exploiting information as a strategic resource, *International Journal of Policy and Information* 8(1):1–8.

KING, W. R., and B. EPSTEIN (1976). Assessing the value of information, *Management Datamatics* 5(4):171–180.

PORTER, M. E. (1980). *Competitive Strategy,* Free Press, New York.

SECTION 3

IMPLEMENTATION OF STRATEGIC SYSTEMS

Articles in this section describe specific, empirical instances of strategic system implementation. For the most part, they involve *strategic uses of information* rather than *strategic uses of information technology* per se. The emphasis is squarely on information.

Roger Levien of the Xerox Corporation describes two different 10-year strategic planning exercises used at Xerox. One followed a traditional staff-executive pattern where skilled staff analyze data and prepare alternatives, and executives discuss and choose among the alternatives. In the second exercise, a more participative strategy was followed. The results were quite different. One conclusion of this chapter is that in order for information systems (manual or automated) to have an impact on strategic planning, senior executive planners themselves must be able to direct that information system rather than be directed by it.

King and Kraemer recount the experience with large-scale econometric models used in the public sector to assist federal policymaking. First they distinguish between systems of *strategic intent* (which are rare) and systems of *strategic potential* (more common). Most real-world strategic systems they find evolve over long periods of time—they were not intended originally to be strategic systems. Implementation of real strategic systems is made difficult by system complexity, the cost of maintenance, and systematic bias.

Henderson and Sifonis describe the components of a strategic information system planning method. They first note that IS and business planners often represent different groups in a corporation and different areas of expertise. Yet their cooperation is vital if information systems are to be used for strategic business purposes. Henderson describes a methodology for IS planning which helps to define IS products and services.

Charles Kriebel concludes this section with a discussion of how strategic information systems are actually accounted for in business firms. Strategic systems—perhaps much more than traditional systems—are unlikely to show returns on investment within the traditional American executive's time frame (e.g., one to three years). Given the penchant of American executives to evaluate investments according to a rigorous ROI criteria, many useful strategic systems will not be built. Kriebel calls for new accounting methods and issues the warning that ultimately the performance of the firm depends on leadership and vision.

CHAPTER 4

Making Strategic Concepts Work

Roger E. Levien
Xerox Corporation

There is an elegant ambiguity in the title of this conference: "Information Technology and Management Strategy" offers at least four different interpretations. First, there is *strategy about information technology.* At the Xerox Corporation the focus of our strategic considerations is on information technology and the way we expect it will evolve in the future. The second meaning is *information technology as a tool for strategy:* How can a corporation use information technology to gain competitive advantage? A third meaning is *information technology in support of strategy making*—the use of computer models to assist in the process of strategic decision making. A fourth meaning concerns not the technology of information, but rather, the use of *information in the making of strategy.* Of those four, the one I intend to discuss is the last. I am going to talk about the way strategy is made, the role that information plays in making strategy, and how to get appropriate information to strategic decision makers at the right time and in the right form.

Consequently, what I say will have relatively little to do with information technology in the usual meaning of bits and bytes, hard disks and floppies, and more to do with information technology in a softer sense; how information is controlled and shaped so as to enable effective strategic decisions to be made. If there is a spectrum of content at this conference, my talk lies at the end of the spectrum most distant from technology.

I have discovered that I am the only one on the agenda, except perhaps for Alan Kay, who is not a card-carrying academic. Consequently, what I will say will have to rely for its credibility on being based on experience in an actual business environment.

My experience before I joined Xerox was in dealing with governmental decision making during 20 years at two think tanks: The RAND Corporation, and an international research organization, IIASA. I remember vividly how my colleagues and I at those organizations used to imagine what it would be like to deal with private enterprise, where—we believed—there would be a small number of clear objectives and precise goals, the data would be plentiful and accurate, and decisions would be made and implemented speedily.

I certainly do not want to disillusion this group of close students of business, but I will say that I found my experience at RAND more useful than I thought it would be; even in private enterprise the reality is decision making under considerable uncertainty. As a planner, in public or in private enterprise, one must help decision makers to interpret uncertain information, to understand the unclear alternatives they face, and in some way or another, to identify the possible consequences of selecting each of those alternatives.

STRATEGIC PLANNING AT XEROX

Let me illustrate the role of information in strategic planning by describing what we have been doing at Xerox over the last few years. I shall characterize and contrast two approaches that we have taken and illustrate what we have learned from them.

The problem that we addressed arose from our view that relevant strategic concepts and information were not available to senior management when strategic decisions were taken. Now, I do not mean to say that decisions were taken in ignorance. Clearly, a goodly amount of information was available, but not of the right kinds nor as much as those of us with an academic background would consider necessary for a fully satisfactory decision to be taken.

The first approach, which we called a "strategic reconnaissance," was named "Xerox '92" because of its 10-year perspective. It was done using a *systems analytical approach,* based on the techniques that I had absorbed during my 20 years at RAND and IIASA. The second project that I shall describe was named "Xerox '95"; it began in 1985 using a participative approach based on the insights of organizational psychology and marketing. The difference between the systems analytical and participative approaches and their different degrees of influence is the central theme of the balance of my remarks.

XEROX IN 1982

In order to understand the two projects, some background information on Xerox is necessary. The primary business of Xerox in 1982 was reprographics (a formal word for copying), although there were other information products as well. Xerox was moving toward the domain of "office systems," a phrase that covered a broad range of products and opportunities. As you know, Xerox is a global corporation, although it operates through Rank Xerox in Europe and Fuji Xerox in Japan. Its outposts are everywhere; however, the majority of its business is in North America (about 50 percent). The revenues in 1982 were on the order of $8 billion.

In 1981, Xerox decided to decentralize—to move away from a highly centralized form of management to a strategic business unit (SBU) or, rather, a pseudo-SBU structure. Unlike true SBUs, which have total business control from design through sales, the Xerox SBU's do not do the sales and marketing on a geographic basis. However, they do have full product planning and production responsibility and define the marketing approach on a worldwide basis; they transfer the product to the operating units, who are responsible for marketing, selling, and servicing it in their geographic regions. The idea was to move business responsibility from the corporate to the SBU level, which could be closer and more responsive to the marketplace. As a part of the decentralization, corporate marketing and manufacturing were eliminated and the corporate planning staff was changed to a corporate strategy office. The intent was that the corporate strategy office would not now be concerned with the day-to-day operations of the operating units and SBUs, but rather would devote its attention to the longer term and the overall portfolio balance of the corporation.

I joined the corporate strategy office in 1982, just after it was created. Because I arrived with no specific knowledge of Xerox, but a lot of experience of long-term planning, I felt that I could make my best contribution by addressing the problems that adequate information about the *longer-term external environment* was not available to senior management when strategic decisions were taken. I would not say that this was a problem that was commonly perceived within the organization. Rather, I volunteered to solve it and was given the support of my superiors to do so. There is an open environment at Xerox and a willingness to listen to new ideas.

A STRATEGIC RECONNAISSANCE

The approach I adopted was one that would be familiar to anybody who has spent a day and a half at RAND. I formed a small, interdisciplinary staff group to perform an analysis. The objectives were to be defined by top management, after which the staff group would identify some possible environmental scenarios. Finally, the staff would look at some alternative strategies and examine them in the light of the scenarios. The comparative results would be organized in a careful and substantial report. Senior management would be appreciative; making new and better decisions based on the better information we had given them.

This is the paradigm, in the crudest terms, that we set out to follow. How did we go about doing it? Because we were not in a position to define real strategic choices for the corporation, we said that we would "fly out over the horizon" to do a reconnaissance of the strategic terrain 10 years ahead—a "strategic reconnaissance." This would be a "high flyover" to get the lay of the land, not a detailed mapping.

The staff group comprised a number of people from around the corporation who, as it turned out, had been doing pieces of such a reconnaissance already. There was market research, competitive analysis, and economic analysis under way, as well as various technology studies, but these data had never been put together so that they made sense. The strategic management question we addressed, therefore, was not "what computer hardware should you use to bring information to top management?" but rather, "how do you gather the information and organize and analyze it so that it might be useful to top management?"

We worked for about seven months and produced a 10-chapter report. After those seven months, I briefed senior management on our findings. It was on December 17, just before the Christmas holidays. As they came into the room, each found a 300-page loose-leaf notebook on the desk, wrapped in a red or green ribbon—I was feeling a little spritely at that point and I thought I would give the presentation a light touch. But 300 pages cannot be given a light touch under any circumstances!

Senior management very patiently listened to my two-hour briefing on what the external world might be like and the consequences for Xerox—and they were very receptive to the concept. They took the books home for the holidays, but I do not know if they read them; we did not give quizzes.

We then had a series of top-management presentations. As it turned out, Xerox had planned a senior management education program for 1983. The most senior 250 people in the company came together in 12 week-long sessions, 20 or so at a time. Over the course of that year, I was able to have the first morning of each of those weeks to brief them on the results of the strategic reconnaissance. Consequently, a lot of information about the future external environment reached the management core of the corporation.

What was the information we provided? The first piece of good advice I had received was from a consultant. He said: "If you are going to do this, for heaven's sake focus your work on a question that top management will be interested in. Don't do a view of the world that's very general—the kind you can get from a book. That kind of general information is not going to be useful. Take a specific question of importance to management and focus on that." The question I chose to focus on, because it was something I thought I could do, was: "How might Xerox revenues grow over the coming decade?" I also chose a particular revenue target. Now I was looking at the external environment not in terms of general themes and trends, but to answer a specific question: "Can the revenues of the corporation reach a specific target?"

We used a very simple conceptual scheme. What we wanted to do was to find the size of the market that Xerox would be dealing with and then say how much of that market would be accessible to Xerox. First, we had to look at the economic trends: Would the environment be one of growth in the world economy or one of shrinkage? We had to understand the demographics: How many office workers are there going to be? How many factory workers? How will that affect the demand for Xerox products? What social forces will be at work—for example, the move to work at home through the use of information technology or the greater participation of women in the work force—and how will they affect the market? Finally, and most important from our perspective: What will be the technological changes that will affect the character of the marketplace and its size?

In general terms, then, these four factors—economics, demography, social forces, and technology—can be brought together to form the basis for some estimate of the size of the market, which is the playing field that Xerox has for the future. However, we will not be the only ones who are playing. There are two other very important forces. One is the government. Especially outside the United States, the government imposes many different barriers to market access or to market penetration. Corporations have to pay a toll of some kind when they come across most borders. They have to meet certain standards, or criteria. In every country outside the United States, there are some requirements that have to be met in order to get access to the marketplace. How will those requirements shave the size of the market available to Xerox? Again, it is possible to draw upon our information to modify gross market sizes to reflect the impact of government policies on the size of the market accessible to Xerox.

Finally, there is the competition; they are not going to let you have it all to yourself. So we also have to estimate the way the competition will evolve. Our equation was: *markets* minus or plus *government policy* ("plus" because in some countries the governments can be helpful) minus *competition* (competitors are never helpful) equals *revenues*. It was that simple scheme that we tried to execute. There were no massive models. We did not have 10,000 simultaneous equations—we did not have that much data, nor did we have any sense of what the equations could be. There were some models of market size, but the data were very crude indeed. What we produced was a book that had chapters on each of these subjects. The end users never saw a model. Rather, they saw information about, for example, economic alternatives and about demographics. In the latter case, we tried to focus the attention of top management on the fact that demographic change will be important for our industry, just as it is for consumer products. It is not just economic numbers or just technology that affects the size of the market for office equipment, but the number of people in offices: it is where low-cost labor is located for factories; it is the percentage of women in the work force.

The key analytical tool that we used was a simple matrix. You can think of it as a spreadsheet. The matrix comprised columns representing the key regions—North America, Europe, Latin America, Asia—and the rows represented Xerox's key product classes, such as copiers, printers, workstations, etc. We then tried to estimate the 1988 and 1992 sizes of the market. What is the total size of the market? How is it divided by product class and region? What is its growth rate? The matrix showed in detail, on a worldwide product-by-product basis, what the market size and growth were expected to be. Then we put down the best current data we had for 1982: market size and growth, current market share, and the share of our revenues that came from each of those markets. Then we estimated the difficulty of increasing Xerox's share in each of those markets. How competitive is it? What is the government's role? Is the technology changing? And so on. Then we made several different estimates about the rate of growth of our revenues in each of those markets. Could we maintain our market share? Could we increase our share? Would we decrease our share? Using those and different estimates, we projected several alternative patterns of revenues. Then we drew the strategic implications that these reconnaissance sorties suggested.

RESULTS OF XEROX '92

A number of important concepts came from Xerox '92. The first was the idea of doing 10-year environmental scenarios, which had not been done in such a way at Xerox previously. We developed the idea of "alternative revenue possibility paths," spelling out different paths the company could follow to gain its targeted revenues, and we drew a series of strategic implications. What markets look most promising? Where were there specific difficulties? We proposed no strategies at that time—just strategic implications. We tried explicitly to deal with uncertainty by using ranges for all numbers. All of these were accepted concepts that I brought with me from the world of the think tank.

What was the impact of all of this? I think we could say with near certainty that information about the longer-term external environment was no longer unavailable to senior management. In fact, we forced the entire management core (over 250 persons) to sit for a half-day and listen to us describe the longer-term external environment; and many of them even read the 300 pages, or at least the 70-page summary, that we also provided. In that sense, the effort was certainly a success. To be specific, many people said "what a great piece of work; marvelous; never have seen something like that before"—and then they put it in their bookcases, never to look at it again. All the nice things were said, but—truth be told—the bottom line is that there was little discernible impact on Xerox's strategic decision making.

I had my fling, I did a nice job, and I was proud of it. There were some ideas that came out of the project that people used and may have in an indirect way affected some strategic decisions, but mainly it was an add-on to the strategic process. It was not embedded in it. It did not have a continuing impact. Therefore, when we came to 1985 and I assumed the position of vice-president in charge of the corporate strategy office, I resolved, along with some of the people working with me, that we were going to do better the next time around. That is what I am going to tell you about now.

XEROX 1985

In 1985, the business, the markets, and the revenues were remarkably like they were in 1982. We still remained decentralized, but a number of important things had happened. Dave Kearns had been designated as the chairman of the board and chief executive officer. He and his senior management team had taken time together to spell out a shared vision for the company in the future. Kearns had worked very hard to build a spirit of teamwork among his senior dozen or so managers—"teamwork" in the sense that while he was the final decision maker, he wanted an environment in which those senior managers participated in the decision-making process as much as appropriate.

And there were some significant changes under way in the culture of Xerox, a company that had grown up rapidly in a market it created and served singlehandedly for over a decade. The world for Xerox had changed significantly in the 1970s because of the entry and rapid growth of competition. In the 1980s, Xerox started benchmarking the competition and setting competitive performance as the goal. It determined that it must meet or beat competitive performance in quality, cost, and service of its products. Every function Xerox performed and each product it made was benchmarked against the competition.

Second, Xerox wanted to increase the extent of employee involvement. Learning from the Japanese, employee involvement teams were set up throughout the company. These and other ideas were merged into a major program called "Leadership Through Quality." The fundamental notion was that every group identified its customers (internal or external), defined their requirements, and then set out using problem-solving techniques to solve the problems that kept it from meeting those requirements. Generally, employee involvement teams were used as the vehicle for problem solving. Throughout the company the concept of active involvement in decision making was emphasized.

Third, we put in place a planning process that had one-, three-, and five-year horizons. This new process was simpler than earlier approaches because it had a bottom-up quality. Let me explain what I mean.

As I noted earlier, Xerox has product-oriented strategic business units that pass their products to the operating units, which market, sell, and distribute the products in their areas. Xerox, therefore, is a matrix organization, with the SBUs playing the primary role.

Each year the SBUs submit five-year strategic plans; these are integrated at the March Consolidated Strategy Review (CSR) to see how they add up to a corporate plan. Top management reviews this portfolio and decides its overall financial and strategic targets and what needs to be cut or filled in to achieve them. Top management is assisted in this process by the corporate strategy office and by corporate finance.

The targets and specific direction then go to the SBUs, who make the necessary changes in their plans and pass the resultant guidance along to the operating units. They, in turn, prepare three-year business plans. These business plans are then consolidated and compared with the strategy that was decided at the Consolidated Strategy Review. The Consolidated Plan Review where this is done is held in June. It produces the targets and guidance to both the SBUs and the operating units to develop their one-year operating plans for the next year. These are consolidated and approved or revised in November.

This is the planning process. Notice that it operates bottom-up from the operating units and SBUs. Whatever guidance and decisions are made at the top management level have, until recently, been limited to the financial areas. What we felt was missing in this process was a mechanism for top-down strategic planning—what we now call a corporate strategy. In its absence, it was unclear what overall direction these bottom-up plans were supposed to serve.

XEROX '95—TOWARD A CORPORATE STRATEGY

In Xerox '95 we have worked with top management to develop a corporate strategy, a strategy that would provide the top-down guidance to direct the resources of the SBUs and operating units. One of the things we had undertaken in designing the planning process was the simplification of planning documentation. We introduced a "Business Resource Management Statement" (BRMS). Instead of a thick plan coming from each of the SBUs, we asked them to give us four pages to describe their strategy. The top page gave us their mission, their market, their general strategic approach, their performance for the last five years, and what they expected for the next five years in terms of return on assets (ROA), cash, revenue, and profits. The second page gave us information on the attractiveness of their industry. How large is it? How competitive is it? What are the customers' requirements? What are Xerox strengths relative to this industry? What are the critical success factors in the industry and for Xerox to succeed in that industry? The third page asked the general manager to give his or her assessment. We asked them what the critical external and internal events are that have to occur for their plan to achieve its objective, the changes from their last plan, and to give us an overall assessment. The fourth page was a financial summary. Through this format, we tried to focus the discussion with top management on strategic issues, not on the day-to-day management of SBUs. We tried to help the SBUs to organize their plan in strategic terms, not to swamp corporate headquarters with data.

It so happens that (appropriately to the topic of this conference) we use advanced computer and communication technologies—Xerox 6085 workstations with Viewpoint software and a worldwide network of local Ethernets—to transmit these strategic plans from around the world into corporate headquarters almost instantaneously. But the advanced technology is not what accounts for its success. What is important is the paring down of the information to the absolute essence and its structuring so that corporate management could concentrate on the essence of the strategic issues.

The issue we faced in 1985 was that the corporate strategy was not sharp, or clear, or explicit enough to support focused strategic decisions. And we discovered as we went along that top management world views were quite different. Even among the top 15 people, there were large differences about such critical matters as the future of paper in the office. Is there going to be paper in offices in the year 2000, or will offices be paperless as some pundits predict? That is extremely important to Xerox since we make most of our money putting marks on paper. Yet top management world views on these fundamental issues differed and those differences naturally led to different strategic preferences. That means that a key requirement of the strategic planning process was to create a shared awareness and agreement about the world among our senior management.

PARTICIPATIVE APPROACH

This time, instead of an analytical approach based on staff work, we chose a participative approach. Since in the end we have to convince top management, the easiest way to do so is to engage them from the very beginning in the process. We decided, therefore, that top management would be the performer or the explicit guide of all steps of the analysis. We in the staff would be there to gather and present information in concise form and to structure the meetings. In other words, we were the information technology. We were the devices to bring together the information and provide it to top management. Top management would work together to understand the environment, identify the alternatives, and then to make recommendations to the chief executive officer.

THE PROCESS

Participative decision making can go just so far in a corporation. In the end, the CEO decided, just as everybody understood he would do from the beginning. But his decision is quite a different decision when everybody responsible for implementing it has been through the process of understanding the choices and has had a chance to let his or her views be known.

The process we followed began with three major meetings held off site in which the 15 top managers of the corporation participated. The first two were one-day meetings and the third was a three-day meeting. We did not exactly lock the doors, but the meetings were away from the office, and they were carefully scheduled. Managers knew that decisions had to be made by the end. In order to engage people very early, we interviewed each of them before the first meeting, asking them questions about their long-term goals for Xerox—what they felt were the critical success factors, critical environmental factors, and so on. We analyzed their replies and gave back to the group (with all of the identifications removed) the summary results of these interviews so that everybody knew the range of opinions, although they did not know who held what opinion.

At the first meeting, we discussed measures of success, long-term goals, and critical success factors. What came out of this discussion was a sense of where we wanted to be in the long term and an awareness that there was quite a range of opinion within the senior management group.

Before the second meeting, we circulated a questionnaire asking questions about criteria of success, strategic preferences, and so on. At the meeting, we went through the results (again without revealing identities) and showed top management that there was a range of questions on which everybody agreed, but there was also a range of topics on which there was quite a range of opinion. That led us to ask: What are the alternatives we face?

Looking at the responses to the questionnaires, we identified four strategic preference clusters. We used the clusters to draft very rough definitions of four alternative strategies. In the afternoon of the first day, we set up four teams and asked each team to spell out one strategy in some detail. After a few hours of separate work, each group came back and reported on its strategy to the others. We agreed at the end of the meeting that these were four realistic, but quite different, strategic alternatives for Xerox that needed to be spelled out in more detail and analyzed.

Before the next meeting we (the staff) spelled out these alternatives in some detail, using a format derived from the Business Resource Management Statement, with which senior management was familiar. At the final meeting, we spent time talking about the different views of the world and presented detailed analyses of the four alternatives. At the end of the two days, each of the 16 members was asked to give his recommendations to the chief executive. On Sunday morning, the third day, the CEO said: "Having listened to all of you, here is the way we are going to go." His choice became the strategy of the corporation.

ANALYSIS

We found that there were four critical elements in our strategic choices. First, what is the future of the office; how is it going to evolve? Second, how is electronic technology going to affect the office? Third, what is the future of the document within the office? Is it going to be paper or electronic, how will people use it, where will it be filed? Fourth, what is the nature of the organization in which offices will exist, and how will information technology be used by those organizations?

We concentrated the attention of top management on these issues. We got agreement among them on a perspective of the world and the charter to keep monitoring that view of the world and to report to them if our perceptions change.

In considering the four strategic alternatives, we used the Hollywood notion of "high concept." A high concept is a one-sentence summary of a play or movie. "Great white shark terrorizes beach community"—you already know the movie. Well, we used a similar high concept to describe each of the four overall strategies. The basic notion we had was that of a *strategy space.* Four strategies in our view bounded the range of possibilities that Xerox had: a very aggressive market growth strategy; a very aggressive profit strategy; a product niche strategy; and a marketing and service strategy.

Real strategies probably lie within a space bounded by these four extreme cases and our first task is to decide where in this space we want to be in 10 years. What is the target point? The second task is to determine how to get there. What is the path from our current location to the target point?

RESULTS

Finally, what were the concepts that came from Xerox '95? First, we embedded into the planning process of the Corporation the notion of a 10-year corporate strategy. We introduced the notion of strategic alternatives and the space of strategic alternatives. We convinced senior management that we need a view of the world on which to base our strategy. We continually monitor the world view and the strategy.

We now have a 20-page document in which that is characterized. Xerox '95 sharpened explicit corporate strategy. It has had an impact on strategic decision making, providing the top-down guidance on the basis of which strategic choices were made among the bottom-up plans presented during the 1986 and 1987 planning cycles.

Let me return to the problem we faced: How can top management be provided with the information required for strategic management? I believe that relevant strategic concepts and information can be made available to and indeed will be used by top management *if the approach is active participation* by top management throughout the strategic process. Staff should not do the planning for senior management, but rather should provide the support for top management to plan. Staff should, in the end, act as *the information system* that will enable top management to make more effective decisions.

The information technology that we used most extensively in Xerox '95 was primitive: pen, pencil, paper, and a few terminals, but basically, people defining and structuring information to help and assist the decision maker. Rather than bombard decision makers with thick notebooks, with lots of data, most of which is irrelevant, we tried consistently to focus on the small number of things that are really critical and that affect strategic choice. Much of what we did was dependent on the particular environment at Xerox and David Kearns's management style. Nevertheless, I believe that the underlying participative approach to strategic management would be useful to and successful in many other enterprises, both private and public.

CHAPTER 5

Implementation of Strategic Information Systems

John Leslie King
Kenneth L. Kraemer
University of California, Irvine

It would be strategically foolish to design an information system if there were no hope of implementing it successfully. Implementing routine information systems for operations and management can be difficult; implementing strategic information systems can be more so. Our purpose here is to examine the important issues in implementing strategic information systems. We will not concentrate on prescriptions about implementing information systems generally. These prescriptions are useful for implementing strategic information systems, but there are many such prescriptions and they are discussed at length elsewhere [see, for example, IBM (1984), King (1978), Kriebel (1968), McLean and Soden (1977)]. Instead, we focus on those characteristics of implementation that are tied to the task of building systems with more than passing strategic significance.

We have divided our presentation into three parts. The first deals with the concept of strategic information systems itself and identifies some implementation difficulties entailed in the concept. The second reviews some factors that have proved important in implementation of complex information systems for decision making under uncertainty, which is the root task in strategic planning. The third summarizes the lessons learned from experience with implementing these decisions, and provides some advice.

IMPLEMENTABILITY

Problems of Definition

The first task in implementing something is to determine exactly what is to be implemented and whether it is implementable. Here we encounter our first problem with strategic information systems. A precise definition of the concept is lacking. It is relatively new to the literature, and not a great deal has been written on the subject. But from the literature we see two definitions that are often jumbled together, although they are different in important ways.

One definition describes information systems that are designed from the start for strategic purposes. In the business context this generally means competition in classical markets where information is free, transaction costs are zero, and there are no governmental constraints on free trade (Porter, 1979, 1980, 1985; Wiseman, 1985). In such cases, "strategic information systems" would be those that that are designed to bring about product or service differentiation, cost reduction in production for competitive purposes, or concentration on a particular market niche. In a slightly broader organizational context, it could mean any enhancement of an organization's productive and sustaining links with its environment (Chaffee, 1985). Strategic information systems here could be part of military strategy, intraorganizational political strategy, or electoral strategy. In either case, these are *information systems of strategic intent*.

The second definition describes information systems that are designed mainly for operational and management purposes, but that can be useful in strategically important ways. Examples would include airline reservation systems that help fill the seats, but that also can be used to direct customers toward particular choices of carrier and service, or on-line order-entry systems located in customers' offices to reduce sales costs, but that also help lock customers into buying from a specific supplier. These are information systems with strategic potential.

Much of the literature on strategic information systems begins with discussion of information systems with strategic potential. The authors advise the intelligent systems manager to identify and capitalize on this potential. But the authors generally turn their attention at the end of their articles toward developing information systems of strategic intent. For example, Ives and Learmouth begin their informative 1984 paper on information systems as competitive weapons with a discussion of the nature of competition and some models for classifying information systems applications in competitive terms. They then discuss a set of information systems that appear to have been built mainly for operations or management purposes, but that were put to work toward strategic ends. Up to this point they have been discussing information systems with strategic potential that happened to deliver on that potential. As they state quite clearly, the models they review are useful mainly for description and not for planning information systems of strategic intent. In the last paragraph of the paper, however, they imply that their CRLC (customer resource life cycle) model does facilitate identification of strategic targets that information systems can be targeted toward. At the last moment the paper's focus has moved away from identifying features of systems that might have strategic applicability and toward identifying strategic targets for information systems application.*

Problems of Implementability

We make this distinction between systems of strategic intent and systems with strategic potential because the implementation of the two kinds of systems have very different levels of difficulty. We make three observations in support of this position.

First, information systems of strategic intent are more difficult to build because there is not much practical experience to go on. As the newness of the subject suggests, there are few information systems of strategic intent in existence. Most systems labeled in the literature as strategic are systems with strategic potential that have been promoted as examples of systems of strategic intent. Even so, systems with strategic potential tend to be large scale, geographically dispersed, and built up over a long time period. For example, the American Airlines Sabre system was started in the 1950s as a nationwide reservation for the American Airlines agents in large cities. It was extended

*This tendency is seen in many articles on the subject of strategic information systems, including Cash and Konsynski (1985), McFarlan (1984), Rackoff, et al. (1985), Beath and Ives (1986), and Porter and Millar, (1985). This is not a criticism of these articles; strategic information systems is a new concept that has not been completely thought through (Treacy, 1985; Bakopoulos and Treacy, 1985). Corporate strategy is akin to the subject of leadership, which is inherently difficult to define and study. Everyone knows what good leadership and shrewd strategy are after the fact, when the consequences of the leadership and strategy are obvious. No one quite agrees on how great leaders and strategists are made or how they do what they do. In the same way, everyone agrees after the fact when a given application of information systems technology has produced a competitive advantage for a firm, but translating that observation into action that produces more such examples is a different matter altogether. The approach to technology and its application to strategic objectives from the IS perspective differ in interesting ways from the more general issue of technology as a factor in organizational strategy (Frohman, 1982).

beyond large cities to travel agencies in the 1960s. It was opened to service other airlines, including international carriers, in the 1970s. And the technology platform of the Sabre system was used as the base for American's frequent business flyer program, which was extended to individual customers in the 1980s. Thus, the system's scope and scale expanded over a period of three decades.

It is unclear at what point in its history the Sabre system was recognized by its designers or managers as "strategic," but it is likely to have been relatively recently—the late 1970s or early 1980s. More important, this example illustrates why information systems of strategic intent are rare. Few organizations can make the long-term, large-scale commitments to their construction up front, even if they could conceive and plan such systems. Moreover, the current popularity of the "prototyping" approach to systems development suggests that the conventional wisdom in systems development is moving *away* from the kind of grand designs required by systems of strategic intent.

Second, information systems of strategic intent are more costly to build, not only in direct costs, but in opportunity costs and risk. The strategic context of most organizations is not well understood in precise terms sufficient to facilitate construction of formal (e.g., game theoretic) systems where outcomes can be predicted with considerable certainty. Therefore, information systems of strategic intent potentially involve great opportunity costs and risk. It is for this reason that even if conceived as strategic to begin with, they are likely to be built as systems with known operational and management payoffs, while their strategic aspects remain a hope. Also, because of their tremendous costs, it is likely that most efforts to build strategic information systems capability will pay off *only* if these systems produce substantial practical benefits in operations or management in addition to whatever strategic advantage they provide.

At a more general level, building a good system of strategic intent is as difficult as doing good strategic planning—in both cases one is aiming at a moving target. As military strategists well know, few battle plans survive contact with the enemy. Battles are won through fighting, not through planning. Good planning sets up the opportunities and helps reduce exposure to loss, but no amount of planning will guarantee success. Even if one can imagine every contingency, the cost of planning for each remains tremendous. The only pure examples of systems of strategic intent are found in national defense. The costs of these systems are very high. These systems suffer from the same problems noted above, but the tasks they perform are so important that even brute-force methods such as running the systems under all plausible conditions are often employed.

Third, information systems of strategic intent are more difficult to evaluate. Under the best of circumstances it is very difficult to assess the impact of such systems, or the reasons why they perform successfully or unsuccessfully. The outcomes that really matter—organizational performance and success—are the result of many forces that cannot be controlled for, and the true role of the system in the success can be hard to prove. Post-hoc accounts, no matter how carefully done, do not usually provide complete information on the impacts or processes by which the impacts result. The identification of particular systems as "strategic" has thus far been made by outsiders looking for examples to fit the concept. When they come upon an interesting and successful example of an information system used for strategic advantage, they immediately assume that the designers who built it did so to achieve the strategic effects that have materialized. Few of the systems held up as examples were conceptualized as strategic systems by their designers. Detailed accounts of how these systems came to be, when they are provided at all, seldom address whether the systems were designed as strategic from the beginning, as opposed to having been turned from operational or management systems into tools important for strategy.

At this point we are uncertain about the promise of systems with strategic intent. Using just the experiences of corporations, there is not much that can be said about successful implementation of such systems. As time passes and more organizations gain experience with applying information systems to strategically important tasks, this might change. In the meantime, efforts to improve the concepts behind and the techniques for information systems in strategy should continue. There are clearly opportunities in applying information systems to the strategic realm because information systems can make a difference in many facets of organizational life. It is likely, however, that systems with strategic potential will continue to be more frequent contributors to organizational life in the near term. Comparatively speaking, we know more about what they are, how to plan and build them, and how to evaluate their costs and payoffs.

IMPLEMENTATION

Admonitions from Prior Experience

The general study of the implementation of innovations is a curious field. Our colleagues who edit journals in operations research, management science, and computing complain that they receive few submissions of good articles on implementation. Apparently, many people in the field prefer to build things and leave the issues of implementation to others. This is analogous to the situation in software development, where skilled analysts and programmers want to develop systems but do not wish to maintain them. Mobilizing the talent to affect a successful implementation can be difficult. Among other things, system failures are often attributed to implementation when the real culprits were poor design or simply a bad idea in the first place.

This underscores some important points about implementation. The first is that implementation of strategic information systems is incremental. Generally, strategic information systems as well as traditional systems are built in increments that evolve, through redevelopment, toward more integrated, comprehensive, and sophisticated systems. Even when conceived as a single system, a strategic information system must be built in pieces. An effective strategy for implementation, therefore, is to develop sequential sets of systems, each of which builds upon preceding established capacities (Kraemer and Perry, 1979; Perry and Kraemer, 1979). The second point about implementation is that for systems that will serve complex tasks such as strategic planning, there is no real line between development and implementation. Such systems are always in development. They are first under development in the process of being created; later they are in development because they have been implemented and need to be changed. The standard view of the system "life cycle," which draws sharp distinction between development and implementation, does not describe these systems too well. In a sense these systems never leave the requirements definition phase. Life cycle models, and the admonitions built on them for implementing routine systems, do not provide sufficient guidance for creation of strategic information systems that continually evolve to take advantage of new opportunities. We do better to investigate successful systems that are in a perpetual state of tuning and refinement, because the objective of implementation is to keep systems competitive—not just get them running.

Fortunately, for our purposes, there are some systems strategic intent that provide real-world guidance for implementation. We focus here on the successful implementation of highly complex decision support systems for national policymaking in the controversial areas of economic policy and social welfare program planning (King, 1984a,b; Kraemer, et al., 1987; Kraemer and King, 1986, 1987). These systems are clearly used for strategic purposes, often in ways more overt than many corporate cases would reveal. We spent several years investigating two such systems. The first was the Data Resources, Incorporated econometric modeling system of the U.S. economy. This system is a collection of complex models (some with more than 1000 equations), very large data bases of economic time series data, and computing and telecommunications systems to facilitate analysis. At the time of our study in the early to mid-1980s, the DRI system was the most extensively used modeling system in the federal government, with at least 50 using agencies. The other was a duo of very similar microanalytic simulation models, the Urban Institute's TRIM (TRansfer Income Model) and Mathematica Policy Research's MATH (Micro Analysis of Transfer to Households), that are used to evaluate the behavior and effects of huge income transfer entitlement programs (e.g., social security, AFDC, medicare). Judging simply by extent of routine use, these are among the most successfully implemented planning systems in history.*

Our studies revealed that the implementation of computerized planning models among federal agencies was dependent on "supply push" factors associated with modelers' work, coupled with the agencies' "demand pull" for quantitative analyses (Roessner, 1979; Tornatzky et al., 1983). The intensely felt need for modeling that began in the mid-1960s greatly facilitated the process of implementing major models for use in actual policy analysis. Within the welfare policy area, there has been long-standing concern over the effectiveness of the nation's large, complex, and costly income transfer programs, including welfare, health care, and social security. This was combined with the desires of modelers to make microsimulation modeling an attractive tool for social welfare policy analysis in Washington. Within the macroeconomic policy arena, a persistent desire to gain greater control over the economy has maintained constant pressure to improve technical understanding of the economy. This has combined with academic research that built a broad base of macroeconomic models, federal government action to create major economic data bases, and the advent of advanced computing capability. In both cases the result was the creation of service organizations to provide modeling services to an enthusiastic audience of policymakers and their aides.

These successful cases of model use arose and survived at the intersection of two forces: persistent political and policy need for tools to aid analysis and decision making; and an aggressive community of model builders who had secured access to government data bases and the requisite computing capability to permit analysis. They were implemented successfully in spite of the difficulties of their implementation because there was terrific need for them, and there were people willing to build and maintain them. There also was a strong academic interest among social scientists and economists in building and maintaining the models, so an invisible community of academics was available to provide ongoing "no cost" support to the enterprises.

Barriers to Implementation

These systems contain many of the same problems that confront information systems of strategic intent. They are difficult to implement in the first

*The DRI model is used by the Congressional Budget Office, the Troika (the economic analysis group made up of the Council of Economic Advisers, the Treasury Department, and the Office of Management and Budget), and many of the federal mission agencies. It is also used extensively by commercial and industrial organizations. The TRIM and math models are used by the Congressional Budget Office, the Office of Management and Budget, and the mission agencies involved in transfer programs (HHS, Agriculture, Labor).

place, so the fact of their success is testimony to the fact that difficult obstacles can be overcome. The obstacles themselves need to be carefully considered, however, since the special conditions leading to success in the case of policy models might not apply to more operationally oriented strategic systems in companies. We identified four that appear to be most important.

The first obstacle is the complexity of the required system. The more complex the computerized modeling system is for the real user, the greater the difficulties associated with implementation and use. Complex systems are harder than simpler systems to learn to use, and pose numerous operational problems. Many users of these systems cite model complexity as the largest obstacle in the implementation process. This is a serious problem because there appears to be a drift toward complexity in the creation of such models. Systems technicians tend to be attracted to complex models. Complexity appears to be necessary for capturing the behavior of complex problems. Also, complex systems simply look impressive to others, which increases the influence of those who use them. Unfortunately, systems that are difficult to use or comprehend tend to be used less often by the decision makers who matter. This was a serious problem with the TRIM and MATH models, but the essential need for the models ensured their use by a select clientele of staff specialists who provided a critical technical infrastructure for the decision makers. In contrast, the complex DRI systems provided sophisticated software routines to make their use relatively less troublesome. DRI was in a position to sell sophisticated capabilities to system novices, and its large base of subscribers provided the millions of dollars that made the development and maintenance of that capability possible.

The second problem is that of generating and maintaining the necessary data for the models. It is extremely difficult to obtain sufficiently large and representative samples of data to permit detailed modeling of complex phenomena such as economic and human behavior (the subjects of the TRIM, MATH, and DRI systems). The only reason these systems could work at all was the tremendous investment in data collection infrastructure by the federal government. Even the DRI system depended in large part on receiving its ongoing time series data from government data collection agencies. All government agency users agreed they could never amass the data provided by DRI, and DRI acknowledged it could not afford to create such data bases on its own. Only the existence of government agencies like the Bureau of Labor Statistics and the Bureau of Economic Analysis that are specifically charged with collecting such data made DRI's system possible.

A third problem arises from the real or perceived presence of systematic bias in the modeling system or the data bases used. This phenomenon appears in the application of models to politically sensitive issues, of course, but it also appears in nonpolitical situations. This is not a problem that can be solved technically. All models that simulate complex and dynamic systems have relatively low predictive accuracy. Moreover, the results of modeling that are passed back to actors in the system can alter those actors' behavior in ways that artificially confirm or disaffirm the model's forecasts. Everyone involved in decisions cares about the accuracy of the models being used, regardless of whether they understand the difficulties in achieving accuracy. When models fail to produce accurate predictions, confidence in them wanes rapidly.

A final problem in model use is the difficulty in establishing use of model-generated information as relevant to the decision-making process. Analysts tend to be strongly supportive of modeling as a key component in their work, but they are constrained by the political realities faced by policymakers. The objective is not simply to have the "right" answers in the pure sense of the term, but to *win* in political debate. Ambiguity and indecisiveness are marks of weak leadership, so policymakers insist on receiving point estimates rather than range estimates from their models. They want to know what will happen, not what might happen. Explanations of why an estimate is not completely reliable does not remove concern about reliability. Unless a model can provide concrete assistance in answering the questions of interest to the policymaker, the analyst's advice will be ignored and support for modeling in the policymaking organization will break down.*

LESSONS FROM EXPERIENCE

We can draw some lessons from these experiences that are relevant to strategic information systems. To do so we must first identify the similarities between these modeling systems and the concepts of strategic information systems noted earlier. The modeling systems are in some ways more "strategic" than most corporate information systems. They were designed to facilitate the practical task of setting government policy, but that task is itself fraught with competition and controversy stimulated by partisan disputes between Democrats and Republicans, constitutional disputes between Congress and the Executive, and disputes among executive branch mission agencies. Even the most rudimentary operational decisions can become embroiled in strategic fights among these actors. If the primary question is "Can systems like this be implemented successfully?" the answer is yes. But there are critical differences

*We do not want to suggest that policymakers in the federal government are naive or foolish about the complexities of forecasting and strategizing. Most are very cognizant of these. Their concerns arise because a model that produces only rough guesses of what could happen is really no more dependable a seer than any other kind of "expert," and often is less so. Top decision makers are usually surrounded by experts, most of whom have particular biases and opinions that color their judgments. Model use is successful because it helps focus attention on the issues and the measurement of key variables related to the issues. But the models do not do away with the question of who to believe.

between national policymaking as part of the political process and corporate policymaking. In government, programs are funded ideologically and delivered managerially. The ideological vision behind a program need not be reconciled to the program's results. In contrast, most private firms have to reconcile policies with effects rather quickly. Quite often the objective is to obtain maximum strategic advantage out of everything that happens. More practically, in many firms strategy is limited to finding ways to do things differently in order to have immediate advantage over one's closest competitor. In this way, the fundamental strategizing of corporate leaders cannot depend on broad ideologies. Strategy must be a part of the basic managerial and production activities of the organization.

For this reason we believe that information systems in corporations that do not serve ongoing operational or management needs will seldom serve strategic needs, and those that fail to serve operational or management needs will not be around for very long. The strategic goal for information systems designers therefore should be to ensure that operational and management systems are designed and built in a way that facilitates strategic uses if such uses become apparent. This means installing the necessary features of the systems that enable selective change and growth, continuing deployment to varied user groups, and modeling and analysis capability to permit exploitation of inherent system features for strategically important purposes. For example, if an operational system produces data that can be analyzed to identify key customers for targeted marketing, the systems themselves should be designed with the ability to sort records on these characteristics, write reports, and distribute them on-line to appropriate users. This might require DBMS, report generation capability, additional telecommunications, and attention to the user interfaces (especially if the users have varying experience with the technology or the particular system).

When care is taken to provide systems with capabilities that enable users to exploit opportunities for strategically relevant application, it is fair to say that the systems are used strategically. But it is important to note that it is very difficult to *design* systems with a predetermined strategic utility because of the nature of strategy. Most strategic concerns arise in one of two situations, neither of which is easily accommodated into the world of systems design. In one case strategic concerns arise quickly, usually in response to a competitor's move. The objective is to react to that move and, hopefully, position oneself to gain an advantage out of the repositioning. In the other case the strategy is laid out far in advance of the outcomes being planned—indeed, some strategies are designed to prevent the occurrence of particular outcomes.* In

*Military deterrence comes readily to mind and illustrates one of the interesting paradoxes of strategy. The objective in the case of military deterrence is to maintain such a war-ready profile that war does not occur. Of course, this case is quite extreme and not analogous to any business situations we can think of.

either case, operational and management systems cannot be designed to accommodate these characteristics of strategy making for most firms because one would either have to be able to tell the future, or have the resources to accommodate every possible contingency. Operational and management systems can be designed to provide *classes* of capabilities that repeatedly show up as strategically useful, but the presence of such capabilities does not make these systems "strategic."

Systems can be built to facilitate strategic activity, as illustrated by Ives and Learmouths's (1984) reduction of Porter's model of the goals of strategizing (differentiation, cost reduction, and finding a niche), to the simpler notion of maintaining a relationship between the company and its customers and suppliers. The examples of strategic information systems listed in that article are really examples of systems designed to affect changes in the production process that either directly or indirectly cut costs, identify/clarify markets, or make customers happy. All have a direct and obvious connection to the production process. In short, they all contribute in more or less marginal ways to company performance. That contribution to performance is strategically important, given that performance is naturally and appropriately tied to survival, growth, and profitability in market economies. It does not detract from American Hospital Supply's on-line order-entry systems to point out that a major objective behind their development was to cut the costs of serving small hospitals that did not contribute enough business to make it worthwhile to service them with on-site salespeople.* What we are suggesting is that all efforts to improve organizational performance involve use of information. Generally speaking, those with the best information will be the most successful. But this is nothing new: Benjamin Disraeli made a statement to that effect many years ago, and he certainly was not the first person to notice this truth. Improving the organization's ability to glean useful information from existing information systems offers promise for improving performance.

We do not want to imply that building operational and management systems with strategic capabilities is an easy task. It is not. The primary architects of these systems in the past have not been information system professionals. They have been the strategic planners and corporate managers who understand the business, its markets, and its broader environment. For example, the DRI modeling system was designed by two economists who understood the need for economic forecasting by virtue of their previous roles as policymakers and knew how to convert theories into executable models. They knew the relevant data sources needed for the model and how to organize the data to make them readily available in formats needed by policymakers and analysts. Similarly, the architects of MATH and TRIM were policy analysts first and

*This account of the AHS system was provided in a personal discussion with a principal in a consulting firm who was involved with the development of the system.

model builders only secondarily. And the architects of most of the strategic information systems known as classics in business were functional managers and corporate managers, not information system professionals.

This raises the question of our message to information system professionals. The IS professional's role in most instances will be that of catalyst, facilitator, and technical advisor rather than designer. If IS professionals are to become the designers of strategic systems, a role that many are not currently trained to assume, they must develop the ability to think strategically as surrogates for functional and general management in the firm (Kraemer and Dutton, 1979),* and to apprehend those features of the business that are relevant to strategic application of information systems. This extends the well-known and often intractable problem of having system designers see the world through the eyes of users. System designers and other key information system professionals are usually busy just keeping up with the technical features of the systems world, and seldom have time or inclination to become intimately familiar with the domains of the applications they design. Despite these problems, the incorporation of strategic thinking into the system designer's repertory of concerns is *essential* if systems to facilitate strategy are to be built, regardless of whether information system professionals play a narrow or expanded role. We suggest that the organization's commitment to strategic information systems will hinge on the willingness to invest in training designers to incorporate strategic ideas into systems and in training strategic planners and managers in system design.

CONCLUSION

Overall, we believe the best course toward successful implementation of strategic information systems is to concentrate on making the organization itself competitive. This, as Porter (1980) points out, can take many forms, including use of information systems. Exploitation of information systems is likely to produce major strategic opportunities only in two circumstances: when information systems are the organization's primary production technology or when information systems provide unique and new control over the organization's primary production technology that facilitates rapid and flexible response to new market opportunities, changes in demand, or changes

---

*Roles other than "surrogate" for functional and general management are possible. For example, Danziger and Kraemer (1987) suggest the additional roles of consultant, mentor, and broker. We believe the surrogate role is particularly relevant to the design of strategic information systems, but these other roles might be more appropriate in specific instances.

in supply. Few organizations have information systems as the primary production technology. For most, information systems are only a part of the overall production technology mix. Often, they do not provide the primary technological opportunity to make strategically important changes, although they might be critical in making the best use of the organization's primary production technologies. In these cases, again, it is the ubiquitous nature of information systems in the value chain that provides the opportunity for action. Small, incremental changes that yield small, incremental improvements do not look very appealing taken one by one. But in the aggregate, across the value chain, these incremental improvements can add up to a strategic change, and they can also have a synergistic effect on the quality and responsiveness of management and managers.

REFERENCES

BAKOPOULOUS, J. A. Y., and M. E. TREACY (1985). Information technology and corporate strategy: a research perspective, *CISR Working Paper 124*, Sloan School of Management, MIT, Cambridge, Mass.

BEATH, C. M., and B. IVES (1986). Competitive information systems in support of pricing, *MIS Quarterly* 10(1):85–96.

CASH, J. I., Jr., and B. KONSYNSKI (1985). IS redraws competitive boundaries. *Harvard Business Review* (March–April):134–142.

CHAFFEE, E. E. (1987). Three models of strategy, *Academy of Management Review* 10(1):89–98.

DANZIGER, J. N., and K. L. KRAEMER (1987). *People and Computers: The Impacts of Computing on End Users in Organizations,* Columbia University Press, New York.

FROHMAN, A. L. (1982). Technology as a competitive weapon, *Harvard Business Review* (January–February):97–104.

IBM (1984). *Information Systems Planning Guide: Business Systems Planning,* IBM Corporation, New York.

IVES, B., and G. P. LEARMONTH (1984). The information system as a competitive weapon, *Communications of the ACM* 27(12):1193–1201.

KING, J. L. (1984a). Ideology and use of large-scale decision support systems in national economic policymaking, *Systems, Objectives, Solutions* (December).

KING, J. L. (1984b). Successful implementation of large-scale decision support systems: computerized models in U.S. economic policymaking, *Systems, Objectives, Solutions* (May).

KING, W. R. (1978). Strategic planning for MIS, *MIS Quarterly* 2(1):27–37.

KRAEMER, K. L., and W. H. DUTTON (1979). The interests served by technological reform, *Administration and Society* 11(1):80–106.

KRAEMER, K. L., and J. L. KING (1986). Computer-based models for policymaking: uses and impacts in the U.S. federal government, *Operations Research* 34(4):501–502.

KRAEMER, K. L., and J. L. PERRY (1979). The federal push to bring computer applications to local governments, *Public Administration Review* 39(3):260–270.

KRAEMER, K. L., S. DICKHOVEN, S. E. FALLOWS-TIERNEY, and J. L. King (1987). *Datawars: Computerized Models in Federal Policymaking,* Columbia University Press, New York.

KRIEBEL, C. H. (1968). The strategic dimensions of computer systems planning, *Long-Range Planning* (September).

MCFARLAN, F. W. (1984). Information technology changes the way you compete, *Harvard Business Review* (May–June):98–106.

MCLEAN, E. R., and J. V. SODEN (1977). *Strategic Planning for MIS,* Wiley, New York.

PERRY, J. L., and K. L. KRAEMER (1979). *Technological Innovation in American Local Governments,* Pergamon Press, Elmsford, N.Y.

PORTER, M. E. (1979). How competitive forces shape strategy, *Harvard Business Review* 57(2):137–145.

PORTER, M. E. (1980). *Competitive Strategy: Techniques for Analyzing Industries and Competitors,* Free Press, New York.

PORTER, M. E. (1985). *Competitive Advantage: Creating and Sustaining Superior Performance,* Free Press, New York.

PORTER, M. E., and V. E. MILLAR (1985). How information gives you competitive advantage, *Harvard Business Review* (July–August):149–185.

RACKOFF, N., C. WISEMAN, and W. A. ULLRICH (1985). Information systems for competitive advantage: implementation of a planning process, *MIS Quarterly* (9)4:285–294.

ROESSNER, J. D. (1979). Federal technology policy: innovation and problem solving in state and local governments, *Policy Analysis* 5:181–200.

TORNATZKY, L. G., J. D. EVELAND, M. G. BOYLAN, W. A. HETZNER, E. C. JOHNSON, D. ROITMAN, and J. SCHNEIDER (1983). *The process of Technological Innovation,* The National Science Foundation, Washington, D.C.

TREACY, M. E. (1985). Toward a cumulative tradition of research on information technology as a strategic business factor, *Working Paper,* Center for Information Systems Research, Sloan School of Management, MIT, Cambridge, Mass.

WISEMAN, C. (1985). *Strategy and Computers: Information Systems as Competitive Weapons,* Richard D. Irwin, Homewood, Ill.

CHAPTER 6

Middle-Out Strategic Planning: the Value of IS Planning to Business Planning

John C. Henderson
Massachusetts Institute of Technology
John G. Sifonis
Arthur Young & Co.

The potential for using information technology to affect the competitive position of the firm (Cash and Konsynski, 1985; Rockart and Scott Morton, 1984) has served to highlight the importance of effective information systems planning. That is not to say that research on strategic and tactical IS planning is scarce. As the research review on IS planning approaches reveals (Boynton and Zmud, 1985), there are many alternative methodologies. Rather, as the criticality of effectively linking the strategic IS plan to the strategic business plan has increased, the need to better understand the nature of strategic planning, in general, and strategic IS planning, in particular, has also increased. It is now particularly relevant to ask not only how the IS plan guides the resource allocation for the IS function, but also how this activity adds value to the efforts to devise a strategic business plan.

Venkatraman (1985–86) argues the intersection of interest between IS planning and strategic planning stems not only from a common critical assumption (e.g., a belief that planning positively affects the performance of the firm), but also the similarity in the research questions and methodological issues that have been pursued. Similarly, King (1983) has applied a general framework for evaluating strategic planning processes to the specific instance of strategic IS planning. In each of these cases, strategic IS planning is viewed as a purposeful activity in which organization alignment among a variety of goals and levels within the firm is the key element of success.

When viewed from this overall perspective, the commonalities between strategic IS planning and strategic business planning are apparent. Researchers in both disciplines have struggled with at least four major systems issues:

1. Ways to represent the levels of abstraction inherent in the planning and design process
2. Separability (decomposition) and its implications for creating a narrow planning context or frame
3. The need for cooperative behavior among experts
4. The need to recognize explicitly the performance linkage or the value-added contributions of system components

Each of these issues is resolved explicitly or implicitly by any given planning methodology. In this paper we take the position that the linkage between strategic IS planning and strategic business planning is an indirect consequence of meeting an organization need crested by an attempt to pursue strategic goals. Hence linkage is an indirect result and depends on both the effectiveness of IS planning and the preexisting alignment between the organization strategies and structure. Resolving the foregoing issues will shed light on both of these factors.

In the following section we explore each of these issues in the context of both IS and strategic planning. Then we describe a planning approach that focuses on the issues of consistency between levels of abstraction and the validity of a planning context as two major mechanisms to strengthen the linkage between IS and business planning. Although this approach also calls for a value-based business modeling approach, this aspect will receive limited attention. Finally, we summarize the major concepts founded in the proposed approach and suggest areas for future research.

STRATEGIC BUSINESS AND IS PLANNING

Each of the four planning issues identified above has long been the subject of research. At the core of planning and design is the recognition that this process requires the participant to move between multiple levels of abstraction (Henderson, 1987). For example, strategic planning is often envisioned as having three levels: corporate, business, and functional (Hofer and Schendel, 1978, 1979; Venkatraman, 1985–86). Each level reflects varying sets of stakeholders that are affected by or can affect the plan, the extent to which forces external to the firm are explicitly addressed, the extent to which organizational boundaries within the firm are viewed as constraints, and so on. In general, the planning process addresses the overwhelming complexity represented by a large system by decomposing it according to dimensions such as resources (functional), time, or target market segment.

It is not surprising that this decomposition activity is also found in IS planning. The concept of top-down planning and structured analysis emphasizes the need to decompose a complex system systematically into smaller and more concrete representations. The process of defining the subcomponents and prescribing an action plan is viewed as a series of transformations. A design transform is a conceptual or physical change in the design artifact or target system. Thus, the system design life cycle describes the IS design process as a sequence of transformations that moves the designer from an abstract statement of need to a concrete reality of a system that affects the behavior of individuals within the firm. While the IS life cycle has been used primarily to conceptualize design, there is recognition that this activity must be linked to a predesign or planning process.

It has not been particularly useful to focus research on the possibility that there exists an inherent number of transformations that best describe the planning process. IS planning, for example, has been described as having as few as three and more than 12 levels (Pyburn, 1983). More relevant are two basic issues that must be addressed regardless of the granularity of the levels used to describe the planning and design process: internal consistency and external validity. Internal consistency reflects the need to ensure that the actions envisioned at one level are correctly operationalized at lower levels. As Churchman (1971) suggests, planning and design are, at the extreme, the attempt to prescribe a complete and consistent causal model for a system. To do any less creates the possibility that the proposed plan will not ultimately achieve the desired performance. Although planners would not be so bold as to claim they have a complete and consistent causal model of the firm, they nevertheless strive to attain high internal consistency through the multiple levels of planning. Mason and Mitroff (1981) point out that generating a good solution to the wrong problem is particularly prevalent in an ill-structured environment, and hence IS planning must be concerned with the validity of the plan as well as its consistency.

Internal Consistency

The issue of internal consistency has been addressed in two major ways. The dominant focus of most IS planning methodologies is the internal consistency of behavior. The planning process can be viewed as creating a series of means/ends chains that move from abstract concepts of the firm to realization of particular systems and products that affect the behavior of individuals in (hopefully) predicted ways. In a similar vein, Porter (1980) describes strategic planning as creating a set of goals (ends) and policies (means) for an IS perspective, a major contribution of the critical success factor (CSF) planning methodology is the introduction of a means–end relationship between the goals of individuals and their needs (means) for information. Thus, this planning approach does not ask what information you desire to meet your goals. Rather, first those factors (abstract processes) that will most affect your ability to succeed (goal attainment) are identified and then one asks how these behaviors induce desires for information. This means–end linkage has served to create an intermediate design transformation that has proven valuable to the overall IS planning and design process.

The IS planning literature clearly reflects this emphasis on internal consistency of means/ends relationships (behavior). A review of planning methodologies (Boynton and Zmud, 1985) reveals a common thread of attempting to guide the IS planner through the process of creating these interlinked behaviors. A quite similar tradition is found in strategic business planning as reflected by the flow from corporate to business to function planning (Hofer and Schendel, 1978, 1979; Venkatraman, 1985–86).

More recently, the need to achieve consistency in beliefs as well as behaviors has been recognized by strategic planners. In the strategic planning literature, this has been articulated by research on issues such as organizations enacting their environment (Weick, 1979), and perhaps more indirectly, by methodologies emphasizing participatory planning and design (Kolodny and Stjernberg, in press; Markus and Pfeffer, 1983). Mason and Mitroff (1981) formalized the need to explicitly surface underlying assumptions or beliefs in their strategic assumption surfacing technique (SAST). In part, this approach argues that attempts to gain shared assumptions or at least attempts to clarify and perhaps reduce the conflict that revolves around uncertain assumptions is fundamental to the creation of a corporate strategy. While many of the means/ends (behavior-oriented) methodologies attempt to incorporate discussion of assumptions, the SAST methodology is an example of a strategic planning approach that centers on understanding and, hopefully, attaining consistency in assumptions. Behind this effort is a belief that consistency of assumptions will positively affect behavior.

Henderson et al. (1987) combined assumption surfacing and critical success factor analysis to provide a more comprehensive IS planning approach. Mason and Mitroff (1981) and others have applied SAST or variations on this methodology in an IS planning context. In essence, the IS planning field, as well as the strategic planning field, is recognizing that the lack of consistency with respect to critical beliefs or assumptions is creating a fundamental instability in a plan and hence must be explicitly addressed.

External Validity

The focus on beliefs and assumptions highlights a second major issue that must be addressed in the planning process: the frame or planning context. Planning and design research at the individual, group, and organizational levels (Boland, 1978; King, 1983; Leonard-Barton, 1983; Mostow, 1985) recognizes the fundamental dilemma discussed by Churchman (1971). That is, in order to cope with the complexity of systems, the planner must impose limits or boundaries on the system definition. But doing so clearly limits the scope of the planning effort and hence may not incorporate all relevant coproducers of performance. If this is the case, the planner risks designing a system that is flawed, that is, creating a system that solves the wrong problem. How does the planner validate a given context or frame? If one creates a model of the business, surfaces assumptions, and generates internal consistency for both beliefs and behavior, is there not still a risk of a significant methods bias? Or might not all involved in the planning process systematically ignore an aspect of reality that is critical to the success of the strategy? This fundamental epistemological issue is addressed to varying degrees in the strategic planning and IS planning literature. We refer to efforts to ensure correctness of the planning as the attempt to achieve external validity of the planning process.

Mason and Mitroff (1981) used the concepts of alternative inquiring systems to emphasize the need for the IS planner to explicitly consider the mechanisms for guaranteeing validity of a plan or design. The use of dialectics in strategic planning (Churchman, 1971) is an example of an attempt to increase the likelihood that the chosen strategy is robust and valid. Clearly, the reliance of both strategic planning and IS planning on domain experts reflects the need to establish a mechanism to examine the external validity of a plan as well as the internal consistency of that plan. As the planning environment becomes more turbulent (Huber, 1984a), the issue of external validity becomes more critical as well as more problematic.

Additional Planning Issues

A third major issue addressed in the planning literature centers on the need for cooperative behavior in the planning process. The need to tap many sources of expertise and gain a shared commitment is obviously related to the issues of internal consistency and external validity. Although not the focus of this paper, the need to gain cooperation among experts is a major component of most theories of planning and design. The IS planning literature has borrowed heavily from research on change management (Kolodny and Stjernberg, in press), participatory decision making (Leonard-Barton, 1983), and political science (Markus and Pfeffer, 1983) as a basis for prescribing approaches to design. Strategic planning has recognized both the need to access multiple experts for their knowledge base as well as to incorporate key stakeholders in order to achieve consistency and commitment (King and Cleland, 1977; Pyburn, 1983). How the planner seeks to achieve the distribution of participation and influence is a major component of the planning process strategy.

Finally, the issue of performance must be highlighted. Churchman (1971) argues that a system cannot exist without a concept of performance. Clearly, to define planning as a purposeful activity requires the planner to consider the relation of the recommended system to a notion of performance. The source of a concept of performance often is a major distinction between strategic business planning and IS planning. Strategic IS planning normally assumes that the goals have been passed down via the more abstract process of strategic business planning. For example, the notion of enterprise data modeling for information management does not directly address value. This concept calls for the IS planning process to define the set of all data and their relationships relevant to the firm. These data are then stored in an accessible form. Value is created directly when a process is applied to this data resource. Since the data resource assumes no predefined process, the value added to the firm is the flexibility to support any relevant process or application.

Note that in its pure form, this concept does not attempt to value individual data items. Rather, all data are viewed as potentially relevant to some mind. While such a concept has theoretical appeal to the computer scientist, it begs the fundamental issue of establishing value in the strategic planning process. It will be argued here that a value-based model of the business should drive the view of the data resource. While such an approach cannot guarantee a complete and consistent data model for the firm, it directly addresses the need for strategic planning to attempt to maximize system performance. That is, strategic business planning, as well as strategic IS planning, must focus on those processes and beliefs that will most affect desired performance. Such an effort has its risks, but if done well, should provide clear returns.

In the following section we describe a strategic IS planning approach and its relationship to a strategic business planning process. While the issues of internal consistency and external validity are the main focus, the need for a value orientation is discussed. The requirements for cooperative behavior among experts are left implicit in this discussion.

A STRATEGIC IS PLANNING APPROACH

Strategic Business Plan

Figure 6-1 depicts the proposed planning methodology. The planning process is viewed as an attempt to create an internally consistent and externally valid series of means–ends chains. The sequences of means–ends

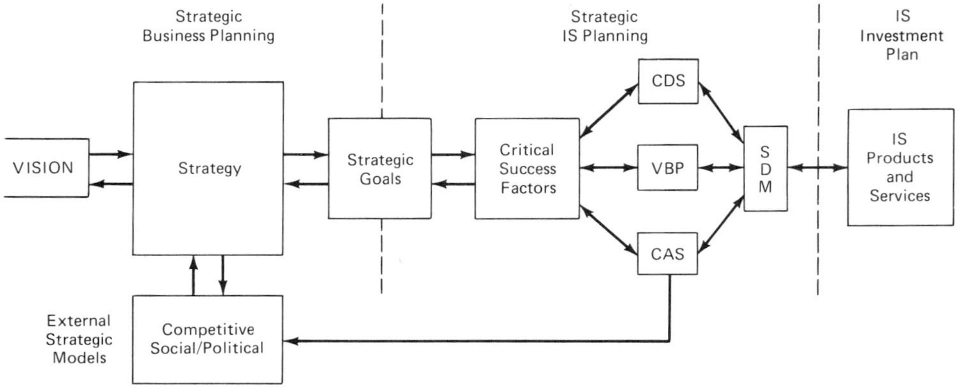

Figure 6–1 The proposed planning methodology.

relationships become increasingly concrete or product/service specific. The planning process is initiated by establishing a vision and strategy relationship. The vision is somewhat analogous to the traditional business planning concepts of mission. It is a futuristic picture of the organization and its environmental surroundings. Strategy is then the macro-level articulation of the means to achieve this vision. It is macro in that it reflects the direction and magnitude of efforts in particular markets and the criticality of various organizational resources to these efforts. In this context, two planning techniques can be used to create this strategy: critical success factors (Rockart, 1979) and strategic assumption surfacing (Mason and Mitroff, 1981). Critical success factors (CSFs) are those things that must be done well in order to reach a vision and successfully implement a strategy. They are the triggers to generate discussion that leads to the creation of policies. The fundamental notion is that while many processes and behaviors contribute to success, there are, in fact, a limited number that are critical to success. Thus the CSF methodology focuses the planning process on a particular few processes that are fundamental to achieving the vision.

The strength associated with focusing is also the greatest limitation of the CSF methodology. The potential to ignore or miss a critical issue exists, as does the potential for an environmental change creating a new set of CSFs. The strategic assumption surfacing (SAS) process offers a means to validate the current set of CSFs as well as to provide a basis to monitor the ongoing appropriateness of this set of CSFs. Assumptions are core beliefs held by the organization about the business environment, organization, competitors, and other relevant stakeholders. These beliefs provide a common basis for action within the organization and often are the source of selective bias about strategy. For example, shared beliefs about potential entrants into a market, even if unstated, may significantly affect the nature of the CSFs associated with a particular strategy. The assumption surfacing process developed by Mason and Mitroff (1981) links assumptions to stakeholders. A stakeholder is an individual or an organization that is affected by or can affect the success of a particular strategy. Thus, the combination of CSFs and SAS provides the key behaviors (processes), beliefs (assumptions), and stakeholds (entities) that are incorporated into a given strategy. It is important to note that each of these methodologies influences the creation of strategy; they do not determine strategy. The actual strategy emerges from the individuals engaged in the process. The explicit use of assumption surfacing and CSFs help to create debate as well as to provide a means to generate internal consistency and assess external validity.

Strategic IS Plan

As indicated in Figure 1, a consequence of identifying the CSFs, assumptions, and stakeholders at the business strategy level is the formation of goals that will be acted on by the organization. The vision/strategy/(means–ends) relationship thus provides the context or frame of reference for a subsequent and more specific means–ends relationship. This phenomenon has been recognized by users of the CSF methodology. Rockart (1979) and others point out the hierarchical relationships that emerge in an organizational CSF planning process. That is, the CSFs for the executive management team often become specific goals for organizational subunits. Henderson et al. (1987) note, however, that while consistency between levels of means–end relationship is one requirement for internal consistency for the planning process, internal consistency must also exist among the critical beliefs or assumptions. As will be discussed, the proposed methodology builds on the need to establish consistency for both behavior and beliefs as one mechanism to ensure that the strategic information systems plan is appropriately linked to the strategic business plan.

The strategic IS planning process uses the goals established in the strategic business planning process to provide an overall frame or planning context. These goals serve the equivalent role in the IS planning process as the vision did in the business planning process. However, the CSFs derived from these goals provide a basis to develop four IS planning products: critical decision set (CDS), value-based process (VBP) model, critical assumption set (CAS), and the strategic data model (SDM).

IS Planning Products. The six objects sets—(1) goals, (2) critical success factors, (3) critical assumption sets, (4) critical decision set, (5) value-based processes, and (6) strategic data model—are viewed as the primary products of a strategic IS planning effort. The CSFs provide the value focus for the strategic planning effort. Their formal use helps to ensure that the debate during the planning process will be directed to policies and behaviors that are critical to the successful achievement of the firm's strategy. It is important to note that the CSF methodology culls individual interviews from a broad range of strategy stakeholders as a means to establish these critical factors. Focus groups or other mechanisms are often used to validate and gain consistency and commitment among these stakeholders with respect to these CSFs. The point is that CSFs are elicited at the level of resource managers charged with attaining specific strategic goals. In this sense, they represent a more concrete behavior than the broader organizational CSFs generated during the strategic business planning effort. Of course, the consistency between the CSFs at these two levels is a traditional measure of the internal consistency of behaviors for the planning process.

The three object sets linked directly to the CSF are (1) the critical decision set (CDS), (2) value-based processes (VBPs), and (3) the critical assumption set (CAS). The CSFs are used as a planning context to help ensure that the elements in each of these sets are value focused. The critical decision set (CDS) are those decision processes that will most affect one or more CSFs. For example, if a CSF is to retain highly skilled employees, a CDS may be the promotion decision, hiring decision, or perhaps the job assignment decision. The goal is to identify a subset of critical decisions from the set of all possible decision processes in the firm. This effort serves both to identify potential products for investment (i.e., DSS) and to help establish those data that are strategically important to the firm.

The value-based processes set recognizes that achievement of the CSF will ultimately rely on critical business processes performing efficiently and effectively. CSF can be viewed as a means to focus a business modeling process. Thus, while recognizing the need to understand functional processes and their relationship to the data resource, the proposal methodology uses the term "value-based processes" to emphasize a focused process model that captures the strategically important processes and their interrelationships.

This concept is similar to the notion of value-added processes described by Porter (1980). However, there is not necessarily the concept of a sequential or value-added flow. Rather, a recognition that each function or process is tightly linked to a CSF, and therefore effective management of it will add strategic value to the firm.

This IS planning product provides two major contributions. First, it provides a monitoring and control perspective to potential sets of IS products and services. This is the traditional domain of the MIS system. Second, it provides a direct link to the existing application base. Since many of the existing systems were developed along functional process lines, the inclusion of this object set will help IS planners to assess their strategy in the context of the existing IS asset base. This aspect will be critical to the ability to translate the strategic IS plan into an effective development action plan.

The critical assumption set (CAS) are those assumptions that underlie the CSFs. They are the reasons why the planners and stakeholders believe the

CSFs are valid. They serve to identify a critical IS product, executive support systems (ESS). Here, ESS are defined as information systems used to monitor and analyze critical assumptions. Since these assumptions often involve beliefs about the environment, thereby requiring external data, the CAS offers a significantly different and important view of the strategic data model.

Finally, the sixth product emerging from the IS strategic planning process is the strategic data model. The representational form of the SDM is often an entity relation data model and appears similar to the enterprise model currently advocated by those pursuing a data resource management strategy. However, the strategic data model differs in a critical way. There is no attempt to create a complete and consistent data model. Rather, the focus is on identifying the significant value-added data classes and how they relate. The strategic data model provides two services. First, it facilitates the coordination of investments across a range of management support systems (e.g., DSS, MIS, and ESS). Second, it establishes where to focus initial efforts to more effectively manage the data resource. Each of these services is of increasing importance to the effective management of the IS function.

Consistency and Validity. Beyond identifying the products of a strategic IS planning effort, the proposed methodology addresses two related issues: the internal consistency and external validity of the planning process.

The internal consistency of the planning process should be considered for both beliefs and behaviors. Since this strategic IS planning approach explicitly elicits both CSFs and critical assumptions, the internal consistency of the planning process can be directly assessed. The strategic business plan is generated from a macro planning context and often involves individuals who will not directly participate in resource planning. As a result, inconsistency in either behaviors or beliefs is likely. The approach taken here allows the planner to recognize the source of these inconsistencies and to focus attention on resolving them.

External validity addresses the potential that a given planning process may omit or incorrectly address relevant factors. An externally valid planning process is one that does not suffer significantly from the collective bias of those involved in the process. The CAS is used as data that can be assessed using multiple external competitive or social models of the firm. The results of this assessment are fed back to the strategic business planner. Assessment that indicates significant omission could result from two sources: (1) the strategic business plan is itself invalid, or (2) the strategic business plan was not communicated or interpreted at the lower levels in an appropriate way.

While many external models could be used to assess the external validity of the IS planning, the two important classes include a competitive model of the firm and a social/political model of the firm. For example, the CAS can be used to determine if the plan addressed each of the five competitive forces described by Porter (1980). A social/political checklist (i.e., social–political factors that could reflect the successful execution of strategy) could also be used to assess the extent to which the IS strategy is sensitive to critical social/political trends such as demographic shifts, trends in deregulation, life-style shifts, and so on.

A final note is appropriate at this juncture. The CSF and SAS methodologies contribute a structure for producing the components or products of a strategic planning exercise. The specific techniques used to implement these methodologies are a related but distinct issue. For example, a dialectic process is often used as a technique to generate and test assumptions. In contrast, individual interviews followed by focus groups is a common technique for generating and testing CSFs. Henderson et al. (1987) use a structure group technique, the nominal group technique, to generate assumptions. The point is that any specific methodology, as with the proposed method illustrated in Figure 1, may be enacted via a variety of specific planning process techniques. A discussion of how to choose such a technique is beyond the scope of this paper but should ultimately consider the unique trade-offs or cost benefits associated with a given technique.

SUMMARY

We have attempted to define the components of a strategic IS planning effort and show how the products of this effort can serve three purposes:

1. Provide a context for defining the products and services to be delivered by the information systems function
2. Provide a basis to establish an internal consistency of a planning process for both behavior and beliefs
3. Provide a basis to assess the external validity of the planning process

An important issue relates to the selection of an external model(s) to be used in the validity assessment. Why choose Porter's (1980) model over other strategy competitive frameworks? As Churchman (1971) argues, the answer to this question is to use multiple models, not attempt to find a single universal model. For example, one could assess the validity of the IS strategy from a social–political perspective rather than a competitive perspective. To what extent do the assumptions address changing social–demographic trends in their markets? Have they explicitly considered possible regulatory or deregulation trends? Even if the strategic planning process effectively used techniques such as stakeholder analysis to minimize the risk of omitting a key issue, poor communication of the strategic business plan is still possible. Thus a validity assessment is still warranted.

A final issue relates the appropriate level of effort committed to the strategic business planning process prior to the strategic resource planning effort. As discussed in planning methodologies such as adaptive design, the level of effort for strategic business planning in the initial iteration should be sufficient to identify highly leveraged opportunities or possibilities that will lead to failure. These should be accomplished with an interest to iterate. In contrast, the traditional top-down strategic planning process often assumes little organizational learning and little iteration as a result of the strategic planning process. The position taken here is that many organizations are facing significantly new opportunities and threats that are due, in large part, to new information technology. As a result, a strategic planning process that emphasizes learning, and hence focuses iterative feedback and validity checks, is an effective strategic planning approach. This middle-out approach to strategic planning can be facilitated by approaches that generate data on both beliefs and behaviors. In particular, the information on critical assumptions offers a basis to operationalize such a middle-out planning process.

REFERENCES

BOYNTON, ANDREW C., and ROBERT W. ZMUD (1985). A critical assessment of information technology planning literature, *Working Paper,* University of North Carolina, Chapel Hill, N.C.

CASH, JAMES I., JR., and BENN R. KONSYNSKI (1985). IS redraws competitive boundaries, *Harvard Business Review* 63(2):134–142.

CHURCHMAN, C. WEST (1971). *The Design of Inquiring Systems: Basic Concepts of Systems and Organization,* Basic Books, New York.

Davis, Randall, and Reid G. Smith (1983). Negotiation as a metaphor for distributed problem solving, *Artificial Intelligence* 20:63–109.

DE BRABANDER, B., and G. THIERS (1984). Successful information system development in relation to situational factors which affect effective communication between MIS-users and EDP-specialists, *Management Science* 30:137–155.

DELBECQ, ANDRE L., ANDREW H. VAN DE VEN, and DAVID H. GUSTAFSON (1975). *Group Techniques for Program Planning: A Guide to Nominal Group and Delphi Processes,* Scott, Foresman, Glenview, Ill.

HENDERSON, JOHN C. (1987). Managing the IS design environment: a research framework. *Working Paper 158,* Center for Information Systems Research, Sloan School of Management, MIT, Cambridge, Mass.

HENDERSON, JOHN C., JOHN F. ROCKART, and JOHN G. SIFONIS (1987). integrating management support systems into strategic information systems planning, *Journal of Management Information Systems* 4(1):5–24.

HOFER, CHARLES W., and DAN SCHENDEL (1978). *Strategy Formulation: Analytical Concepts*, West Publishing, St. Paul, Minn.

HOFER, CHARLES W., and DAN SCHENDEL, eds. (1979). *Strategic Management: A New View of Business Policy and Planning*, Little, Brown, Boston.

HUBER, GEORGE P. (1984a). The nature and design of post-industrial organizations, *Management Science* 30:928–951.

HUBER, GEORGE P. (1984b). Issues in the design of group decision support systems, *MIS Quarterly* 8:195–204.

KING, WILLIAM R. (1983). Evaluating strategic planning systems, *Strategic Management Journal* 4:263–277.

KING, WILLIAM R., and DAVID I. CLELAND (1977). Information for more effective strategic planning, *Long Range Planning* 10:59–64.

KOCHAN, THOMAS A. (1980). Collective bargaining and organizational behavior research, in Barry M. Staw and Larry C. Cummings (eds.), *Research in Organizational Behavior*, JAI Press, Greenwich, Conn., pp. 129–176.

KOLODNY, HARVEY, and TORBJORN STJERNBERG (in press). The change process in innovative work designs: new design and redesign in Sweden, Canada and the U.S.A., *Journal of Applied Behavioral Science*.

LEONARD-BARTON, D. (1983). Introducing production innovation into an organization: structured methods for producing computer software, *Working Paper 103*, Center for Information Systems Research, Sloan School of Management, MIT, Cambridge, Mass.

MARKUS, M. LYNNE, and JEFFREY PFEFFER (1983). Power and the design and implementation of accounting and control systems, *Accounting, Organizations and Society* 8:205–218.

MASON, RICHARD O., and IAN I. MITROFF (1981). *Challenging Strategic Planning Assumptions: Theory, Cases, and Techniques*, Wiley, New York.

MILLER, JAMES G. (1972). Living systems: the organization, *Behavioral Science* 17:1–182.

MOSTOW, JACK (1985). Toward better models of the design process, *The AI Magazine* (Spring):44–57.

PORTER, MICHAEL E. (1980). *Competitive Strategy: Techniques for Analyzing Industries and Competitors*, Free Press, New York.

PYBURN, PHILIP J. (1983). Linking the MIS plan with corporate strategy: an exploratory study, *MIS Quarterly* 7(2):1–14.

ROCKART, JOHN F. (1979). Chief executives define their own data needs, *Harvard Business Review* 57(2):81–93.

ROCKART, JOHN F., and M. S. SCOTT MORTON (1984). Implications of changes in information technology for corporate strategy, *Interfaces* 14(1):84–95.

THOMAS, JOHN C., and JOHN M. CARROLL (1979). Psychological study of design, *Design Studies* 1:5–11.

VANGUNDY, ARTHUR B. (1981). *Techniques of Structured Problem Solving,* Van Nostrand Reinhold, New York.

VENKATRAMAN, N. (1985–86). Research on MIS planning: some guidelines from strategic planning research, *Journal of Management Information Systems* 2(3):65–77.

WALTON, RICHARD E. (1966). Theory of conflict in lateral organizational relationships, in J. R. Lawrence (ed.), *Operational Research and the Social Sciences,* Tavistock Publications, London, pp. 409–426.

WEICK, KARL E. (1979). *The Social Psychology of Organizing,* 2nd ed., Addison-Wesley, Reading, Mass.

CHAPTER 7

Understanding the Strategic Investment in Information Technology

Charles H. Kriebel
Carnegie Mellon University

INTRODUCTION AND OVERVIEW

Early this year in an address to the Society of Automotive Engineers, Robert Anderson, the CEO of Rockwell International, warned that organizations and institutions lag in their ability to understand and exploit the potential for dramatic change presented by information technology (Anderson, 1986):

> We are entering the information age with a lack of information.
> For example, statistics on domestic auto sales do not show how much of those cars actually came from outside the U.S.
> But domestic sales are a component of the index of leading indicators. So when sales are up is that good news for future domestic growth?
> Or is it really good news for Brazil, Korea and Singapore? (p. 21)
> In an era when money moves at the speed of an electronic blip we need to find ways to improve our Government response time. (p. 25)

Forecasting new developments in information technology (IT) is not a key barrier in our understanding because implementation lags invention and development. To "discover" what is coming in IT development, visit university and industrial computer science research laboratories to see the next generation. But understanding the question of strategic impact is another matter; this requires recognizing, and in some cases creating, *innovation.*

Historically, IT has been an enabler and not the driver of new application opportunities, for example, in MIS business and organizational support systems, such as electronic mail, DSS, and ESS. In an article in the November–December 1985 issue of the *Harvard Business Review,* Salerno poses the question: Whatever happened to the computer revolution? Part of her thesis is that the enthusiasm for IT vacillates in cycles is due to vendor pressure, oversell, and "me-tooism" without a sound and continuing commitment by senior management to investment and maintenance of the assets. The situation is made worse by industry fads and by the often narrow bias of MIS groups to focus on application opportunities that are familiar and doable at low cost. The potential for innovation is lost [see also Bowen (1986)].

Business opportunity is part of the answer, and this necessarily derives from, or is created by, corporate strategy. In a recent book Wiseman (1985a) introduces the concept of "strategic IS vision" as the philosophy or culture for IT at the highest levels to affect organizational fortunes. Top management cannot abdicate their sense of mission and responsibilities to lower-level subordinates; even the "chief information officer" is only an architect and "player" in this process (Canning, 1984).

The current and planned investments in IT throughout the world economy are enormous. In the United States it has been estimated that IS and IT now account for roughly 50% of the new capital investments being made each year by major corporations. This magnitude screams for the attention of the CEO—the potential for waste and lost opportunity is real, and so is the extent of the long-term commitment. Moreover, there is a clear need to *debunk* the belief in some quarters that IT evaluation cannot be done. IT investments must pass management scrutiny and effectiveness criteria as other investment decisions must. But the bias for simple tools and answers in this regard needs rethinking; these are not simple questions and the answers are complex. Data and analytical problems are legion; however, top management cannot simply walk away from the issue—at least not without the risk of their own peril. The issue has been with us for some time, it is becoming increasingly important as investments soar, and we are beginning to see evidence of better ways to deal with it.

In the next section we briefly highlight some of the newer technologies for perspective. We then focus on two aspects of management issues: the concept of strategic impacts and the question of investment evaluation and productivity. Next we describe some "new approaches" to address these issues. The paper concludes with recommendations for enhanced understanding.

TECHNOLOGY HIGHLIGHTS

In the fall of 1971 a colleague phoned me to ask if I would participate in a session he was chairing on MIS at the Spring Joint Computer Conference; my assignment was "the future" (Kriebel, 1972). Two concepts I speculated heavily on in that paper were (1) the "model" of the user held by the system designer with the emphasis moving toward management decision processes, and (2) a distributed environment of technology (I used the phrase "distributed network grid"). I never realized at the time how far the second concept would evolve, so I suppose my imagination failed me. Nonetheless, since technology is not my message here, and others at the conference are covering this aspect, I will touch only briefly on a few highlights—adding, I do not foresee any revolutions during the next five years.

The personal computer will continue to grow in power, capability, and flexibility (e.g., Crecine, 1986) at decreasing cost. The professional user will exploit enhanced memory, files, displays, and portable software, while the casual user will find a more friendly environment in which software "agents" will dispatch arduous details. Portability of PC power will also increase as we move from the "table TV" models to the attaché case/wristwatch variety. The larger personal workstations will be nodes in distributed networks, with most of the processing done by powerful local machines and the nets providing file service and communications, as well as the ability to share local cycles given access. Key descriptors for telecommunications will include competition, private networks, interconnection, standards, and very high bandwidths, as more capacity is delivered to the home as well as the office. Through heavy government support, the supercomputer is coming into its own; however, the multiprocessor revolution of harnessing computers together is also on the upswing at large data and scientific centers(e.g., Brauman, et al., 1986; Dertouzos, 1986). In many large industrials today the key word in system design and development is "architecture," recognizing the basic building blocks and interfaces required to integrate functional subsystems and data bases. This awareness will grow in importance. With less emphasis but no less importance is the recognition that the quality of infrastructure (e.g., supporting services, skills, etc.) determines the quality of the IS delivered. Finally, but not exhaustively, the technology of AI/ES and knowledge bases has received notable attention in the press as well as the trade (e.g., Dreyfus, 1986; Kriebel, 1985). This technology will grow in importance as more software aids (e.g., language shells) and experience develop, although not to the extent that some advocates will claim.

Having touched on some highlights, let me continue the story.

MANAGEMENT ISSUES

In a seminal article published during the summer of 1980, Hayes and Abernathy provided an assessment of the malaise and apparent economic decline in the United States. To quote from their summary (p. 77):

> It would be an oversimplification to assert that the only reason for the decline in U.S. competitiveness is that our managers devote too much attention and energy to using existing resources more efficiently. It would also oversimplify the issue, but to a lesser extent, that it is due to their tendency to neglect technology as a competitive weapon.
>
> Companies cannot become more innovative simply by increasing R&D investments or by conducting more basic research....Senior managers who are less informed about their industry and its confederation of parts suppliers, workers, customers, etc. have less time to consider the long-term implications of their interactions and are more likely to exhibit noninnovative bias in their choices....
>
> The key to long term success—even survival—in business is what it always has been: to invest, to innovate, to lead, to create value where none existed before. Such determination, such striving to excel, requires leaders—*not* just controllers, market analysts and portfolio managers. In our preoccupation with the braking systems and exterior trim, we may have neglected the drive trains of our corporations.

The key words are "innovation" and "leadership." What's the management outlook today?

In cooperation with the Japan Economic Journal, the management consulting firm of Booz, Allen & Hamilton, Inc. (1985) recently published a survey of U.S., European, and Japanese executives on "managing for the year 2000." We summarize some of the results of that survey in the executives' responses below.

- The United States ranked first in new product development; but Japan ranked first in the factors enabling the creation of new products (e.g., applied technology, manufacturing experience, good quality control, and a long-term planning horizon).
- Overriding objective for CEO to achieve while in office:
  —For the United States, 51 percent said to "create shareholder value," followed by 18 percent to make company an industry market leader
  —For Japan 41 percent to secure long-term leadership positions in their markets
  —For Europe, similar to the United States, but also very high was "to perpetuate the enterprise"
- How to effect changes in their corporations:
  —For the United States, reduce costs
  —For Japan, expand and strengthen marketing efforts
  —For Europe, similar to the United States
- Finally, how to achieve diversification and growth:
  —For the United States, through acquisition and mergers, with internal growth a longer-term consideration
  —For Japan, through internal development of new products and markets
  —For Europe, through expansion into international markets.

This capsule portrait from the Booz Allen report gives one picture of the types of management skills likely to be perpetuated in U.S. executive offices through the end of the century, namely: accounting, financial, and legal professional backgrounds with some "universal" managers. Does this mean more of the same and further economic decline in the United States as Hayes and Abernathy forecasted in lieu of a change in orientation? Where are the strategists, the leaders, the innovators? We now turn to a couple of examples (from several) as illustrative of the dominant leadership role from top management and the CEO that IT requires.

Strategic Impacts

In recent history a number of case studies have appeared in the IS management literature as exemplars of how IT can be utilized to achieve strategic competitive advantage for a firm. Although not exhaustive, the listing of firms includes such notable organizations as American Airlines, American Hospital Supply Corporation, Avis, Bank America, Citicorp, Digital Equipment, Dunn & Bradstreet, General Electric, General Motors, McGraw-Hill, McKesson Corp., Merrill Lynch, Texas Instruments, Toyota, and United Airlines, to name several [see, e.g., Anderson (1986), Bowen (1986), Cash and Konsynski (1985), Porter and Millar (1985), Schwartz and Petro (1985), Strassmann (1985a), and Wiseman (1985a,b)]. We will briefly describe one case history, Banc One, based on a discussion by Wiseman (1985a,b). Our choice of this case is that it seems best to illustrate Wiseman's notion of "strategic vision" in terms of the level, duration, and extent of commitment to the vision" by top management, especially the CEO. Without this full commitment it is unlikely that the vision and success would have materialized.

Banc One of Columbus, Ohio, was formed as a multibank-holding company in 1967, by John G. McCoy, the CEO. When he succeeded his father as top executive at City National Bank of Ohio (the forerunner organization) in 1959, the bank had under $150 million in assets and primarily business customers. When McCoy retired after 25 years as CEO in 1984, Banc One had assets in excess of $7 billion, with customers mostly from the consumer sector, and it ranked in the top 10% of the 100 largest U.S. banks on common measures of superior performance. Through a series of mergers, it now controls 22 affiliate banks in Ohio. Its earnings per share and net income have increased for 16 consecutive years, to $83.3 million in 1983 with over 6900 employees and 314 branch offices.

Part of the explanation for this extraordinary success, despite increasing competition from giant commercial bank holding companies and financial–retail behemoths like Sears Roebuck and American Express, can be attributed to McCoy's strategic IS vision, according to Wiseman (1985a). To cope with federal and state regulations on geographical expansion, he established the bank holding company, which permitted acquisition of other banks within state lines. To overcome product development restrictions, his managers searched for new products and services as additional sources of income. Some notable results: In 1966 Banc One introduced the first Bank Americard credit service in the United States outside California. In 1971 with IBM it made the first major test of point-of-sale systems using magnetic cards. In the late 1970s Merrill Lynch contracted with Banc One for data processing and Visa debit card services associated with Merrill's new Cash Management Account. And so on. Most recently, Banc One has entered an agreement with Comp-U-Card, a computerized shopping below-retail buying service. The two have created a new card, "Super Visa," which the bank issues and Compu-U-Card offers as an incentive for new customers.

The results at Banc One are impressive, but so were the vision and commitment, over a 25-year span, by the CEO, McCoy. There is nothing in this case history (or others) which suggests that investments in IT and other technologies were treated lightly by management; it is much more likely that the opposite was true given the vulnerability. But top management had a vision for the future, they invested in it, and created value.

Investment Evaluation and Productivity

In an excellent recent article Robert Kaplan (1986) makes an analogous case in discussing investments in computer-integrated manufacturing (CIM) technology and DCF with the following example (p. 87).

> When the Yamazaki Machinery Company in Japan installed an $18 million flexible manufacturing system, the results were truly startling: a reduction of machines from 68 to 18, in employees from 215 to 12, in the floor space needed for production from 103,000 to 30,000 sq. ft., and in average processing time from 35 days to 1.5! After two years, however, total savings came to only $6.9 million, $3.9 million of which flowed from a one-time cut in inventory. Even if the system continued to produce annual labor savings of $1.5 million for 20 years, the project's return [using these figures] would be less than 10% per year. Since many U.S. companies use hurdle rates of 15% or higher and payback periods of 5 years or less, they would find it hard to justify this investment in new technology—despite its enormous savings in number of employees, floor space, inventory, and throughput times.

What's wrong here? Certainly, the concept of discounting future cash flows is not in error, since receiving $1 today is worth more than receiving $1 at a future date. The difficulty lies in the poor and rote attempt to apply DCF analysis to IT investments and the penchant of financially oriented managers to employ simple mechanical rules for complex decision problems. For example, it is still all too common today for investments in IT to be based primarily on potential labor cost savings in cash flow calculations because such costs are "easy to understand," in contrast to something as vague as "strategic advantage." Kaplan (1986) provides a critical review of the traditional pitfalls and some modern insights into the alternatives.

Another accounting-centered issue regarding IT investments concerns recommended professional practices by the Financial Accounting Standards Board (FASB) in the treatment of costs, particularly those for software development. The costs for purchased hardware and most purchased software (especially if it is bundled with the hardware as a "package") are most commonly capitalized as an investment over some useful life of service and accounted for as such. The treatment of the cost of internally developed software is much more ambiguous, notwithstanding the release of Statement 86 by the FASB in August 1985 (see Young, 1986). The earlier remark that roughly 50 percent of new capital investment each year by the large corporations in the United States is being spent on IT and IS includes software development. However, historically, software development costs have been accounted for as an expense item (i.e., written off in the fiscal year incurred). Despite the clarifying attempt of Statement 86, the latitude in interpretation it provides will probably mean continued ambiguity in the accounting treatment of software development costs. The difficulty this poses for IT investment evaluation is the obvious potential for asymmetric treatment of the costs involved.

In general, standard accounting methods often are not effective in capturing the costs and benefits of IT to facilitate analyses, such as DCF. Examples of major costs that are difficult to assess include the drain on the time of management and professionals for development (their lost productivity), business interruption costs from poor systems, lost business opportunities, and loss of competitive position from overemphasis on cost reduction and control applications. Similarly, benefits that are often hard to quantify include elimination of activities and structural changes in organizations, better information quality, improved decision making and communications, new business opportunities, product or service innovations, increased learning, and greater flexibility (e.g., economies of scope). But it is clear that management is not about to abandon existing cost-accounting systems; in many cases they serve the management functions for which they were designed (i.e., monitoring and control). What is needed is to augment these reservoirs with different breakdowns, new data, and measurements. For example, as Drucker (1985) and others point out, white-collar and management productivity is susceptible to measurement.

However, here again management should be wary of simple "solutions." For example, one hears the question: What is the "right level" of investment in IT for corporations? Or, is there a "right level"? One can also find annual surveys reporting on annual IT and communications budgets as a percentage of revenues, or IT employees as a percentage of the total work force, or people and software costs as a percentage of hardware, and so forth. Are these statistics meaningful? Consider Table 1, which contains data on payroll and assets for seven major food companies in 1982. Given these industry data, what is the "right level" of payroll/employee? Of asset/employee? Of payroll/sales? Considering averages over the data set is as meaningless as the raw data. The point here is Strassmann's argument that each organization is as different from another as an individual's fingerprint. Comparisons of such statistics are useless: They explain nothing and can be misleading. There is no "right level" based on industry data; the opportunity set, background, experience, and management in each firm is unique at this level of aggregation. However, meaningful measures of "management productivity" can be defined and comparisons made at the business unit (or profit center) level, according to Strassman (1985a) and van Nievelt (1984). We return to this argument in the following section.

TABLE 1 Payroll and Assets for Food Companies (1982)

Company	Payroll/Employee	Asset/Employee	Payroll/Sales %
Dart & Kraft	$20,605	$64,230	17
General Mills	13,980	36,730	19
IC Industries	22,130	78,790	28
Bordon	17,215	78,010	14
Hormel	36,485	66,080	19
A.E. Staley	33,220	262,990	9
Swift	25,085	42,880	6

Source: Strassmann (1985b).

NEW APPROACHES

Before mentioning some "new approaches" to IT assessment, it is important to recognize that the methods differ along various dimensions of comparison and hence their utility in applicability will vary. For example, they differ with respect to the level, scale, or scope of analysis; some are at the individual item decision level, while others are strategic. Other general criteria for comparing alternative methods include the following:

- *Validity or relevance:* as an indicator of performance or payoff for management that encompasses important inputs and outputs
- *Practicability:* cost to do, time required to understand and do, people effort involved, data requirements
- *Quantifiability:* numerical, multidimensional, and/or "formula based"
- *Informativeness:* as an aid to decision making or problem identification
- *Generality, transferability, or reliability:* to other parts of the organization

The criteria above are not an exhaustive list, but should serve to guide direction in investigating various methods further, depending on the purpose of inquiry.

At the micro level of individual investment decisions it is fairly certain that the concepts of net present value (NPV) and payback will continue to dominate as mechanisms to guide decision behavior. As noted above, the logic of NPV is perfectly rational; however, its application in DCF analyses is often too simplistic. Our advocacy here is for "enlightened DCF" analysis of isolated investments along the lines proposed by Kaplan (1986). Understanding the basic overall information systems requirements of an organization has led to the concept of "enterprise analysis" (IBM, 1982); two popular approaches in this regard are "business systems planning" (IBM, 1981), which develops an "information architecture" from business processes, and "critical success factors" (Rockart, 1979), which structures an IS plan on industry competitive characteristics. Wiseman (1985b) argues that these and similar approaches focus on two generic information systems categories: management information systems (MIS) and management support systems (MSS), such as decision support systems (DSS) and executive information systems (EIS). In his view there is a need to shift attention toward "strategic information systems" (SIS), where the primary use is "to support or shape the competitive strategy of the organization, its plan for gaining (or maintaining) competitive advantage" (p. 233).

In directing attention of IT opportunity to strategic considerations for the firm, this does not mean simply to develop IS support for strategic planning. It also implies more than coordinating the planning of IS with the activities in corporate strategic planning. The thrust is to incorporate IT as an explicit element of strategy, to change the competitive environment to gain advantage for oneself over competitors. This might involve altering market structures, channels for logistics, or traditional relationships between suppliers and customers [e.g., Cash and Konsynski (1985)]. One approach to develop insights along these lines is to analyze the "value-added chain" for the organization from the receipt of raw materials and services as inputs, through the production/conversion processes in the organization, to the outputs of goods and services, and beyond [e.g., Porter and Millar (1985)].

A related approach with origins in the arithmetic of "value-added" is the concept of "management productivity" or "return on management" developed by Strassmann [see Strassmann (1985a,b) and van Nievelt (1984)]. "Management productivity" is defined as the ratio of "management value added" (which is the sum of gross profits before taxes and "management costs") to "management costs" (which is the sum of management purchased services, management employee costs, and management capital costs). In brief the authors divide the accounting world of organization costs into a dichotomy: those pertaining to "operations" (or direct production) and those pertaining to "management" (or "nonoperations"). The analysis then seeks to determine management's contribution to performance, and the role of IT in terms of other organizational factors, such as "percentage of knowledge workers," "investment intensity," "management vertical integration," and so on.

Other research with a similar objective in isolating the contribution of IT to performance has employed the more traditional mechanics of microeconomic analysis utilizing production function frontiers [see, e.g.,

Stabell (1982), Jonscher (1983), and Loveman (1986)]. Somewhat more recently there has been work which focuses on efficient cost frontiers utilizing the nonparametric technique of "data envelopment analysis" (DEA) [see Banker et al. (1984), Banker (1985), and Chismar and Kriebel (1985)]. The principal difference between the two approaches is that the microeconomic analysis seeks to identify and estimate stable, parametric functional relationships which describe output products in terms of input resources, including IT capital and labor, at the level of the business unit, firm, or industry. In DEA the emphasis is on the management decision unit and the analysis seeks to describe elements on the efficient cost frontier (i.e., best performance) in terms of the input factors that drive or determine cost performance. Said differently, DEA interest centers on the management decision implications, whereas the microeconomic/econometric interest is in the structural/technological description of the production process.

Each of the last three general approaches hold promise for better understanding of the potential contribution and impact of IT to management performance. However, each are highly dependent on the availability and quality of empirical data in the firm to perform the analyses. Setting priorities for identifying, collecting, and organizing these data could be the most productive next step for management.

CONCLUSIONS

In the interest of brevity, we conclude the discussion with a summary statement of the key propositions that have been argued in the text.

- Forecasting future developments in information technology is *not* an issue for management concern.
- *All* capital investments, including IT, require management scrutiny and evaluation.
- New approaches to IT review show promise, but the "strategic IS vision" concept is not an easy commitment.
- Senior management is well aware of this issue, but they must recognize for IT the need for better, relevant, empirical data.
- IT is no substitute for superior management.

The key to understanding strategic opportunities for IT is to deliver superior management: "to invest, to innovate, to lead, to create value where none existed before." For the CEO, leadership and innovation in IT can deliver the *bottom line:* to maximize the long-term value of the firm to the shareholders.

REFERENCES

ANDERSON, ROBERT (1986). Technology outpaces understanding in the information age (commentary), *Financier* X(1):20–25.

BANKER, RAJIV D. (1985). Productivity measurement and management control, Chapter 9 in Paul Kleindorfer (ed.), *The Management of Productivity and Technology in Manufacturing*, Plenum Publishing, New York, pp. 239–257.

BANKER, R. D., A. CHARNES, and W. W. COOPER (1984). Some models for estimating technical and scale inefficiencies in data envelopment analysis, *Management Science* (September):1078–1092.

BOOZ, ALLEN & HAMILTON, Inc. (1985). Nikkei survey: managing for the year 2000, *Outlook* 9:12–32.

BOWEN, WILLIAM (1986). The puny payoff from office computers, *Fortune*, May 26, pp. 20–24.

BRAUMAN, J. I., ET AL. (1986). Computers, *Science*, February 28, pp. 935–978.

CANNING, RICHARD G., ed. (1984). The "chief information officer" role, *EDP Analyzer* (November).

CASH, JAMES I., JR., and BENN R. KONSYNSKI (1985). IS redraws competitive boundaries, *Harvard Business Review* (March–April):134–142.

CHISMAR, W. G., and C. H. KRIEBEL (1985). A method for assessing the economic impact of information technology on organizations, in L. Gallegos, R. Welke, and J. Wetherbe (eds.), *Proceedings of the 6th International Conference on Information Systems*, December 16–18, University of Minnesota Press, Minneapolis, Minn., pp. 45–56.

CRECINE, J. P. (1986). The next generation of personal computers, *Science*, February 28, 935–978.

DERTOUZOS, MICHAEL L. (1986). The multiprocessor revolution: harnessing computers together, *Technology Review* 89(2):44–57.

DREYFUS, HUBERT and STUART DREYFUS (1986). Why computers may never think like people, *Technology Review* (January):42–61.

DRUCKER, PETER F. (1985). How to measure white-collar productivity, *The Wall Street Journal* (Editorial page), November 26.

HAYES, ROBERT H., and WILLIAM T. ABERNATHY (1980). Managing our way to economic decline, *Harvard Business Review* (July–August):67–77.

IBM (1981). *Information Systems Planning Guide: Business Systems Planning*, 3rd ed., IBM Corporation, New York.

IBM (1982). Enterprise analysis, *IBM Systems Journal*, 21(1).

JONSCHER, CHARLES (1983). Information resources and economic productivity, *Information Economics and Policy* 1:13–35.

KAPLAN, ROBERT S. (1986). Must CIM be justified by faith alone? *Harvard Business Review* 64(2):87–97.

KRIEBEL, CHARLES H. (1972). MIS technology: a view of the future, *Spring Joint Computer Conference Proceedings* (40):1173–1180.

KRIEBEL, CHARLES H. (1985). The shaping of management decision science with intelligent technology, *G.S.I.A. Working Paper 33-85-86*, Carnegie–Mellon University, Pittsburgh, Pa.

LOVEMAN, GARY W. (1986). The productivity of information technology capital: an econometric analysis, *Working Paper*, Department of Economics, MIT, Cambridge, Mass.

PORTER, MICHAEL E., and VICTOR E. MILLAR (1985). How information gives you competitive advantage, *Harvard Business Review* (July–August):149–160.

ROCKART, JOHN F. (1979). Chief executives define their own data needs, *Harvard Business Review* (March–April):81–93.

SALERNO, LYNN M. (1985). What happened to the computer revolution? *Harvard Business Review* 63(6):129–138.

SCHWARTZ, R. M., and F. A. PETRO (1985). The role of systems in implementing strategy, *Outlook* (Booz, Allen & Hamilton, Inc.) 9:46–51.

STABELL, C. B. (1982). Office productivity: a microeconomic framework for empirical research, *Office: Technology and People* 1:91–106.

STRASSMANN, PAUL A. (1985a). *Information Payoff: The Transformation of Work in the Electronic Age*, Free Press, New York.

STRASSMANN, PAUL A. (1985b). What executives should know about information payoff, *The Consultant* 3(1):5–6.

VAN NIEVELT, M. C. AUGUSTUS (1984). Management productivity and information technology, *Information Strategy: The Executive's Journal* (Fall):9–15.

WISEMAN, CHARLES (1985a). Strategic vision, *Computerworld*, May 20, In Depth, pp. ID/1-ID/16.

WISEMAN, CHARLES (1985b). *Strategy and Computers*, Dow Jones–Irwin, Homewood, Ill.

YOUNG, ARTHUR (1986). *Software Costs: Implementing the New Accounting Standard*, Arthur Young International, New York, 23 p.

SECTION 4

EMERGING TECHNOLOGIES

The articles represented in this section describe specific technologies that have or are likely to form the basis of major strategic moves: personal computers, telecommunications services, and expert systems. Rather than being monolithic, each technology exhibits unexpected turns and twists, subtleties that must be mastered to understand its full potential.

Martin Elton considers the plethora of discretionary communications services, including audio and video conferencing, messaging and on-line information and transaction services. He indicates the current status of each and probable trends. Then Elton identifies a number of reasons, some practical, some organizational, and some cultural, for particular technologies catching on while others have dissipated.

Dick Mason considers the criteria for applying expert systems to real-world situations. He illustrates these points by describing his experiences in designing an expert system to support law enforcement officers during hostage seizure.

The tensions and contradictions embodied in technology are evident from these papers. Features of technology interact with various environmental factors to either promote or retard its use and acceptance.

CHAPTER 8

Discretionary Telecommunications Services: Prospects and Problems

Martin C. J. Elton
New York University

This paper is concerned with new telecommunications services intended for the business user. We review problems that have been encountered during the last 15 years and comment on prospects for the next few years. Services are defined from the perspective of the user rather than in terms of their technological infrastructure. Thus the services of interest here are electronic mail, teleconferencing and videotex, for example, and not communications satellites, local area networks, or local exchange service.

More specifically, there is a focus on discretionary rather than nondiscretionary services. With the former, prospective users tend to have an element of choice in whether and how to use them. Airline ticketing personnel, however, do not have the same choice with an on-line reservation system, an example of a nondiscretionary service. The distinction is somewhat loose, one reason being that a service may eventually progress from being discretionary to being nondiscretionary. Nevertheless, it serves as a useful reminder that some classes of user, typically managers and professionals, cannot be programmed in the same way that more junior employees can be.

Discretionary services are intended for hands-on use by those who are not specialists in communications or computing. It is envisaged that they will be employed by lay users and, for this reason, perhaps they receive a level of attention in the media which, currently, seems out of proportion to their significance relative to other developments in the field of telecommunications. With a few exceptions, the adoption of such services has fallen far short of expectations. As will be seen, more than lower prices will be required to remedy this situation.

In the next section we raise issues concerning the place of discretionary services in corporate strategy. The sections that follow provide an overview of the more significant services. The final sections of the paper turn to prospects for the next few years.

DISCRETIONARY SERVICES AND STRATEGIC ADVANTAGE

When considering the role of discretionary services in the achievement of major strategic objectives, it is wise to be cautious about short-term possibilities. Admittedly, there have been some dramatic successes. For example, Boeing started using videoconferencing in 1980 when faced with a very tight schedule for the development and testing of its new 757 aircraft. Management attributed success in meeting the schedule to the use of videoconferencing, which it also credited with several million dollars of savings in engineering and manufacturing costs. Such examples, however, are few and far between.

Expectations about new services are too high. The failure rates for new products in relatively well understood consumer markets are substantial. Why should one expect them to be lower in the far less well understood area of communications services? Moreover, one must be realistic about time scales. It took the telephone 70 years to achieve a penetration of 50% of U.S. households; facsimile transmission systems were invented even earlier than the telephone. While the pace of change is much faster now, there is a tendency to exaggerate the probable speed of diffusion of new services.

One reason for such exaggeration is underestimation of the problems of implementation. The new services of interest in this paper require change in users' behavior; such change may be threatening. They frequently also require organizational change. It is well known that in these circumstances implementation is somewhat fraught. What is curious is that in particular cases this is so often overlooked.

A contributing factor is the extravagant hype surrounding new services. The obvious danger is that it produces an overreaction against previously exaggerated expectations. A more pernicious effect is the way it exacerbates more fundamental problems. By encouraging facile approaches to implementation, accompanied by unrealistic time scales and unrealistic budgets, it may make success less likely. In addition, it diverts attention from the very real conceptual and theoretical problems that are inextricably linked to the innovative services.

Basic conceptual understanding of emerging discretionary services is very poor. It is remarkable that proponents can be so far wrong as to position a service that requires content produced in advance as one in which content will be generated spontaneously by users, and vice versa. Yet, as noted below, this is what happened with both video teleconferencing and videotex. (Maybe it is not so remarkable: decades earlier similar mistakes were made with television and the telephone.) Even when this mistake is not made, there is considerable scope for error about how a new service will be used. The Bell System's Picturephone was a case in point.

Two other systematic mistakes are common. One is, from the outside, to see a class of services as far less heterogeneous than it really is. Teleconferencing provides a good example. In many cases it is quite inappropriate to discuss it without acknowledging that the term embraces five or six very different clusters of services. Within each cluster, heterogeneity is substantial as well. The other and more serious mistake is the use of a simplistic substitution framework in prospective or retrospective evaluation of a new discretionary service. Again, teleconferencing provides a straightforward example. Often, the alternative to the use of teleconferencing is not an in-person meeting. Rather it is noncommunication or delayed communication or communication among a less appropriate selection of participants.

Addmittedly, the simplistic substitution perspective provides a very convenient basis for cost-displacement studies—a decidedly dubious criterion for innovation, anyway, as others have pointed out. Part of the trouble is that it distorts reality and undersells a new service. What is more, it points people in the wrong direction when looking for worthwhile applications, encourages them to misperceive communication as a collection of independent events suspended in time, and obscures the value of hybrid combinations of the new and the old.

There is, then, a need for caution when considering new services. Their successful implementation within a corporation is rarely an easy matter. Paradoxically, this increases their strategic relevance. Successes that can be imitated quickly and easily are much less likely to offer a lasting strategic advantage.

There have been some important successes. A comparison with less successful innovations provides some tentative guidance for those who seek strategic advantage. A good start is to focus upon applications. With the possible exception of electronic mail, the introduction of subsequently successful services into corporations was generally driven by intended applications which were accorded a high priority.

As both teletraining and telemarketing show, when considering the role of new telecommunications services, it can be helpful to consider generic applications before individual services. Both these types of application encompass a variety of telecommunications services, some of them quite humble. Both are likely to continue to grow in importance. For example, national expenditure on corporate training is enormous and increasing; comparative evaluation of alternative means of provision is relatively straightforward; and there is plenty of convincing evidence of successful application of several different discretionary services. (The related field of customer training also appears promising.)

Nevertheless, it is possible to find some cases of success associated with service push rather than applications pull. This is illustrated by the experience of a few large corporations, with deep pockets and strong nerves, which successfully introduced in-house point-to-point videoconferencing services.

From a vendor's perspective, Hooper (1985) offers advice which also has important implications for corporate users. He cites a colleague who "believes it to be axiomatic that a successful business is created from a combination of old need and new technology, or new need and old technology. Meeting new needs with a new technology is a risky undertaking." The remark is open to different interpretations, since new technology which appears familiar to the end user would seem to be inherently less risky than new technology which is unfamiliar. Whatever one's interpretation, it does suggest that the surest path between here and there may be a zigzag rather than a straight line. In other words, if one's planning scenario calls for new applications via new technologies, an intermediate stage—involving existing applications via the new technologies, or the new applications via existing technologies—may be advisable, even if, in a limited sense, it is not cost-effective. Although such zigzags are characteristic of technological progress, they frequently appear to be fortuitous rather than intended.

The challenge of new discretionary services is that they relax constraints. They make possible new forms of organization, new ways of delivering services, new services to deliver, and qualitatively different information for management and planning. Such opportunities call for innovations which are often hard to plan and hard to implement; they go far beyond simply installing new services. Life would be easier if the services in question could be justified initially on the basis of displacing costs without requiring much organizational change. Then it would be possible for users to grow accustomed to them before being faced with organizational change. To date, however, this has not been the general rule. We turn next to an overview of individual discretionary services.

PICTURE TELEPHONE

The failure of Picturephone in the early 1970s continues to be instructive. It provided an early indication that the incremental value of a visual channel in person-to-person communication was almost certainly less than had been anticipated. Moreover, there were probably better uses for it than portraying faces. It is interesting to compare the Bell System's design with that of a European manufacturer which advanced as far as field trials in the mid-1970s, but no further. The latter's device was optimized for portraying objects and diagrams, not faces. The concept was modular: the audio terminal was a high-quality loudspeaking telephone with an added keyboard; the display screen could form part of a dumb terminal for access to a mainframe computer; another module comprised a device with a joystick for steering one's way around a remotely located microfiche; and there was a choice of transmission bandwidths so that regular broadcast television definition was possible, too. Another difference was an underlying assumption that the European service would be used initially for communication over short distances, often within a single plant. The point of the comparison is that even then, the design of picture telephones provided more fundamental design choices than was generally realized.

Several small companies are now reintroducing picture telephones to the North American market. Possibly they will enjoy success in certain specialized environments; picture telephones enjoyed a modicum of success in health care and criminal justice systems the last time around. There seems no reason, however, to expect them to be of value for general business communication.

SYNCHRONOUS GROUP COMMUNICATION: TELECONFERENCING

Several very different forms of service are described by the term "teleconferencing":

- *Audio teleconferencing.* Provided that appropriate equipment is available, this is essentially unrestricted as regards the number of participating locations or number of participants per location.
- *Visually enhanced teleconferencing.* This usually employs regular telephone networks for transmission. Enhancements include captured frame television, which provides still television images, and electronic drawing devices, such as graphics tablets.
- *Point-to-point video conferencing.* At present, this is generally restricted to two sites and six to 10 active participants per site. (In some cases other sites can be "patched in" to see the interactions between the two primary sites and to participate in an audio-only mode.)

- *Point-to-multipoint (or one-way) video teleconferencing.* This is essentially one-way, closed-circuit television, generally delivered via satellite, often incorporating an audio return via the telephone network from the receiving sites. Its primary uses relate to public relations, launching new products and training.

A considerable amount of user research was conducted on the first three forms of teleconferencing during the 1970s. A variety of counterintuitive results were obtained, not least that, for some purposes, the effectiveness of the audio-only medium surpasses that of both video teleconferencing and in-person meetings. Significantly, none of the research foresaw the rapid emergence of point-to-multipoint video teleconferencing in the 1980s.

During the last 15 years, there have been substantial improvements in the technology of audio teleconferencing and in its supporting services (e.g., bridging services that replace the conference operator). Nonetheless, progress has been largely user-driven. Today, it is in widespread use for diverse purposes and available indicators suggest that its use and penetration are growing rapidly. Users value its flexibility and convenience, as well as its economy; to see it as a poor person's video teleconferencing is comparable to regarding a family car as a poor person's truck.

Enhanced audio teleconferencing has seen a bewildering amount of product innovation. Nevertheless, it has so far failed to live up to its promise. A contributory factor may have been the expansion of the term "video teleconferencing" to encompass it, with a consequent unjustified raising of expectations. Its best prospects appear to lie in training and in working sessions among dispersed engineers and other professionals, where it has achieved some noteworthy successes. Although it enjoys the flexibility of transmission over the telephone network, it is limited by incompatibility among terminals.

Relative to forecasts, the uptake of point-to-point video teleconferencing has been dismal. This is often attributed to the high cost of private services and to the inconvenience of shared facilities. There is more to it than this, as shown by the frequent failure of on-premises services with insignificant variable costs. Inflexibility is a contributory factor. So, too, may be a misperception of the medium itself: Video is most useful when one wants to add an element of drama to a meeting; in many cases, however, such an element is irrelevant or counterproductive.

Nevertheless, there are two indicators which are modestly encouraging. First, reported sales of codecs over the last two years have shown a sharp increase. Of course, orders are placed well in advance of services opening for business, so this should be seen more as a vote of confidence by users than as evidence that video teleconferencing has arrived at last. Moreover, the orders have been heavily skewed toward the Department of Defense and the defense and areospace industries. Second, it is the author's impression that short-distance services (e.g., between a downtown and a suburban location) have enjoyed a relatively higher success rate. While attention is still focused primarily on long-distance services, this may change, helped, perhaps, by an increasing number of broadband local area networks.

Point-to-multipoint video teleconferencing has enjoyed impressive growth in each of the last four years. With increasing acceptance and falling costs of technology, there has been a shift from the ad hoc use of public facilities, especially hotel chains, to the installation of "networks" on corporate users' premises. Attempts are now starting to broadcast special-interest television programs for reception over these networks.

For many years, teleconferencing was generally seen as a substitute for travel (mainly long-distance travel). It was implicitly assumed that such key variables as timing and mix of participants would be unaffected by a change from an in-person meeting to a teleconference. (This was a rather remarkable misconception given the lip service paid to systems thinking.) One of the most encouraging developments, over the last few years, has been a growing realization of the role of teleconferencing in adding value rather than reducing costs. Its primary impact is in reducing delays in decision making and in the flow of information for other purposes.

ASYNCRONOUS COMMUNICATION: MESSAGING

Asynchronous services for person-to-person and group communication comprise:

- *Electronic mail.* Even when facsimile, telex, and hybrid services involving hand delivery are excluded, there is a bewildering array of different public and private electronic mail services (of the text variety).
- *Computer conferencing.* This cousin of electronic mail is designed with an emphasis on group communication and includes features for search and retrieval of items from the shared, ongoing transcript.
- *Voice mail.* This is the audio version of electronic mail. Since it uses the regular touch-tone telephone instrument as a terminal, standards are not a major issue.

Electronic mail (via packet-switched networks), a service invented by users (albeit computer professionals), has been available in various forms for over 15 years. Although it seems well established, the 1985 market for electronic mail was estimated (by Dataquest) as being only $300 million. Current thinking is that growth has been seriously retarded by lack of standards. A major attack on the problem was initiated at the start of 1986 with the launching of the Corporation for Open Systems.

Computer conferencing has been available since the early 1970s. Despite the zeal of aficionados, its use in the business community is less than modest. It has been the subject of a substantial amount of user research, and various reasons have been suggested for its disappointing progress. They include lack of discipline on the part of users, underestimation of the importance of facilitation, and application to purposes which are of insufficiently high priority for users.

Voice mail is used primarily for intracorporate communication and for communication with outsiders who telephone into the corporation. Although it made a promising start early in the decade, so far the market has fallen far short of projections.

ON-LINE INFORMATION AND TRANSACTION SERVICES: VIDEOTEX

A wide variety of on-line services serve a variety of specialized needs: for stock quotations, legal information, scientific and technical information, and so on. On the whole, they are intended for specialists, so they probably do not fall into the loose category of discretionary services. Videotex, on the other hand, is clearly a discretionary service. In Europe, it has been promoted to both business and residential customers, although the relative emphasis accorded each has lurched dramatically to and fro. In North America, where, significantly, on-line information services are better established, the emphasis has been primarily on the residential market. Nevertheless, here, too, there have been attempts to promote the use of private videotex services within corporations (or within networks comprising corporations and their dealers—e.g., the automobile industry).

By far the most vigorous promotion of videotex has occurred in France. The French government provided funds to make terminals available at no charge to residential subscribers; the PTT no longer provided them with hard-copy telephone directories. Currently, more than 1 3/4 million terminals are in use and the PTT reported that its revenues were in excess of $50 million for the first half of 1986. The revenues of service providers exceeded this amount. In Britain, Prestel, the public videotex service, has been operating at a profit for the last two years. In Sweden, a videotex service for business customers was operating in the black within two years of its launch. It appears that in several European countries there is a market for public and private videotex for business communication.

Also of interest is the extent to which British and French use of public videotex services has been driven by messaging and user-generated content. This was not in the minds of those who originally developed the services; they saw videotex as primarily for retrieval of professionally packaged information products. The profusion of messaging services and on-line information services already established in the United States would appear to place videotex in a less favorable position here.

TECHNOLOGICAL TRENDS

The adoption of new discretionary services by corporations over the next five years or so depends, in part, on improvements in the technology; likely prospects are briefly surveyed in this section. However, some of the other, less predictable factors noted in the following section are likely to be at least as important.

Picture Telephones

New attempts to launch picture telephones have recently started and are continuing. These will be desktop devices, probably employing 56-kbs circuits for transmission and/or operating over local area networks. Some will provide motion video; some will provide only still images; others may be capable of operating in either mode. One of the challenges yet to be met is to combine motion video with acceptable audio on the same 56-kbs circuit.

Teleconferencing

The personal computer will continue to drive change. It will be used for encoding and compressing graphics, whether captured from a camera or generated via a tablet. There will be more new software for allowing dispersed users of personal computers to share what is on one another's screens, while participating in an audio conference. There will be innovation in sharing the same telephone channel for simultaneous voice and data. And there will be some use of personal computers for encoding motion video for video teleconferencing.

Progress can be anticipated in overcoming the problem of providing visual interaction in video teleconferencing among three or more sites, but issues of human factors and transmission costs will provide formidable difficulties. The current trend toward increased compression, hence lower bit rates, will continue, and if the quality is found acceptable, may be particularly helpful in the development of economical services.

Workstations

Continuing innovation in this area is certain. (From the standpoint of communications, there is an overlap with teleconferencing and picture telephones, discussed above.) Important questions surround selection of the package of features to incorporate in any workstation. An emphasis can be expected on user friendliness, which will require due attention to be paid to audio and paper, as well as what appears on the screen. Some specialization can be anticipated in the development of workstations aimed at different professionals: engineers, architects, and so on.

Group Problem Solving

One of the more intriguing areas for innovation is in combined communication and information services to support problem solving within groups. This is another category which overlaps with teleconferencing, but it also involves access to data bases, computing, and electronic messaging, as well as multiscreen information displays. While there are products under development, it is particularly difficult, given the current state of understanding, to speculate about their potential significance in the short term.

Voice Mail and Audio Information Services

Continued improvement in the price/performance ratio of voice mail systems can be foreseen. The area of interactive audio information services is more speculative. Allowing users the choice of synthesized voice for output, which is currently provided in the form of print, is well within the state of the art. (It is already available in some electronic mail services.) Control of such service by means of the telephone keypad is a feature of some "audiotex" services, used, for example, for airline flight information and reservations. There are potential applications of audiotex for intracorporate communication, as well as at the interface with customers.

Videotex

Several major computer companies have introduced products to support in-house videotex services during the last few years. As yet, the majority of the large users of such systems are those companies themselves. If this use of videotex proves more feasible than other North American applications of the concept, the important need that will emerge will concern integration of the videotex systems with pre-existing corporate data bases. Videotex is likely to prove of greater consequence in Europe, where services face less competition from established on-line data bases.

STRUCTURAL AND CONCEPTUAL PROGRESS

Progress in successful application of new discretionary services has been held back by a variety of barriers which will not be overcome by inventing new or improved technologies. Prospects for the next few years depend in large measure on advances in the following areas.

The telecommunications industry is highly fragmented, particularly that part of it which is most concerned with new services. Potential purchasers are confused; they have to assume the burden of integrating systems or paying for an outside integrator; and they are concerned about the survival of potential suppliers. Teleconferencing provides a clear example.

Standards are of crucial importance in telecommunications. There have been recent signs that users' irritation at the absence of standards is provoking a response by the industry. However, it takes time to agree to standards; they are sometimes ignored; and remarkably, even when they have been observed, compatibility has not always ensued.

In the turmoil following divestiture, minimizing the costs of basic voice and data traffic, while maintaining acceptable standards of service, has understandably consumed much of the energy of corporate telecommunications departments. Less time has been available for planning and implementing new services.

Too often the basic concepts underlying new discretionary services have been dangerously flawed. As a consequence, the methodologies used for needs assessment, evaluation, and implementation have proved simplistic. Other factors that will affect the pace at which new services are adopted, such as the state of the national economy, are beyond the scope of this paper.

CONCLUSIONS

In the first stage of adoption of a new service, it is generally used for applications that do not mark a radical departure from established ways of doing things. Typically, these applications do not displace sufficient costs to pay for the service. The first stage, however, is also a time of learning. Conceptual understanding of the potential of the new service grows; trial and error produce improved procedures for implementation; and opportunities for the service to contribute to organizational change or service innovation are recognized.

In the second stage, this learning is put into effect. Applications are more radical or systematically integrated into ways of doing things. At this point it is clear that the service is economically justified, and formal evaluation is generally seen as irrelevant.

New services can fail in different ways. They may never really enter the first stage because they do not offer individual users enough for them to become comfortable. The more they demand of users, the more they must give back. Computer teleconferencing and picture telephones failed in this way.

Alternatively, they may languish in the first stage until funds or patience run out. This may be because the value-added applications are not there or because they are not discovered soon enough. At present, visually enhanced teleconferencing appears to be languishing, although it may be rescued by certain applications in the field of technical training. Electronic mail and point-to-point video teleconferencing may, at last, be close to the transition from the first to the second stage.

Sometimes, in a particular corporation, strong external pressures can enable a service to bypass the first stage; the example of Boeing's very tight schedule for the new 757 aircraft was noted earlier. However, examples of rapid success with new applications of unfamiliar services have been relatively rare.

REFERENCE

HOOPER, R. (1985). Lessons from overseas: the British experience, in M. Greenberger (ed.), *Electronic Publishing Plus*, Knowledge Industry Publications, New York.

CHAPTER 9

An Assessment of Expert System Technologies

Richard O. Mason
Southern Methodist University

THE CHALLENGE

It is estimated that by 1990 U.S. corporations will spend more than $1.5 billion for expert systems and related artificial intelligence–based products. Some of these investments will be highly profitable; others will result in major losses. Success, like that Schlumberger experienced in the DIPMETER Advisor, will depend on several things:

1. The ability to identify problem areas for the effective application of expert systems (ES) methods
2. The ability to select cost-effective hardware and software for applications
3. The ability to elicit expertise and translate it into functioning systems
4. The ability to implement the systems effectively within an organization

These desiderata are rapidly becoming requirements. American

Management Association executive Thomas R. Horton, for example, predicts that by 1990 all large and medium-sized companies and many small ones will find it "absolutely necessary to acquire and use knowledge-based systems" to be competitive. This prediction is driven by two socioeconomic forces.

- The need to improve managerial and white-collar employee productivity (these employees constitute about one-half of the work force today)
- The need to protect companies and agencies against vulnerability to the loss or unavailability of key experts

A third force lurks in the background: the need to meet the challenge of Japan, Germany, France, and Great Britain's fifth-generation computer systems and applications.

These are significant opportunities as well as threats. Expert systems (ESs) can be applied to contemporary social problems such as hostage taking, hijacking, and terrorist actions. ESs may be employed in these areas to help law enforcement officials make better decisions in complex, stressful situations. Richard Vedder and I have recently developed a prototype of such a hostage-taking information and tactical support system we call HIT (Vedder and Mason, 1986). In HIT, knowledge about the appropriate handling of a situation is represented by rules that are embedded in an incident-based decision tree network. Four different decision makers—the initial responding officer(s), the on-scene commander, the SWAT commander, and the hostage negotiation unit commander—each of whom have similar, but distinct, information needs are supported. Their efforts are also coordinated by the system, which functions in real time as the crisis unfolds. The underlying knowledge and advice in HIT is distilled into clear, concise, highly focused dialogue and displayed in a palatable user interface, since frequently only seconds are available for a decision to be made and implemented. Because the hostage taking and hijacking have become pressing challenges to domestic and international crisis management, we believe that HIT has the potential of becoming a socially beneficial application of expert system technology. In the long run, HIT and systems like it may represent more than an opportunity for applying the technology. Implementing them may become a responsibility.

It is for these two reasons, then—economic necessity and social responsibility—that expert systems technology should be considered carefully by the executives who bear these responsibilities. Unbridled optimism and enthusiasm must be tempered by feasibility and tractability. Thus we must examine expert systems applications with decisive criteria. I offer five.

CRITERIA OF APPLICATION

Among the criteria to apply to any area of potential application of expert systems are the following.

Replication of Application

There should be numerous occurrences of situations for which the choice of effective responses depends on a common body of knowledge and expertise. The need for this criterion follows from the quasi-public good nature of knowledge itself. It is very costly and time consuming to produce in the first instance. The marginal or incremental cost of reproducing it, however, is comparatively quite small. This means that as the number of applications increases, the average cost of application decreases. Since the payoff per application tends to remain relatively constant, repeated applications enhance the economic feasibility of a project. This criterion was met by our HIT project. Hostage taking, hijacking, and terrorism, the targets for HIT, have become problems occurring frequently worldwide. The most frequent causes of hostage taking—domestic unrest and armed robberies—are also occurring more frequently (Strentz, 1983).

This criterion is also met by most projects in the three broad areas to which expert systems have been applied: diagnosis, procedures and classification (generally systematic procedures), and policies. Diagnostic programs such as MYCIN, the medical aid; UCCEL's "Terresa" project, which identifies problems with software bugs; or Texas Instruments' 810 Hotline Helper, which ferrets out solutions to 810 printer problems, share the process of deciding the nature of a repetitive disorder by means of a careful examination of the facts.

One area of repeated application of expert knowledge is in the chemical analysis of objects. Application of computer technology to this problem area led to what, by my reconnoitering, was the first expert system. To determine chemical structure a sample of the substance is fragmented by a bombardment of electrons in a mass spectrometer. This produces an array of ions of different masses, data about which can be plotted on a graph. The x axis represents the molecular weight of the ions; the y axis represents the frequency with which they occurred. Prior to DENDRAL, organic chemistry specialists would examine these graphs and related data intensely in order to select from among the thousands of possible isomers the particular molecular structure of bonds and atoms that obtains for the sample substance. Today, DENDRAL and its descendants have become standard tools and are used by chemists throughout the world to help them determine the probable structure of substances.

DENDRAL is a procedural and classificatory system. This type of expertise has proved to be a fruitful one for expert systems applications. PROSPEC-

TOR, for example, does much the same in geology as DENDRAL does in chemistry. It infers geological structure and the possibility of mineral deposits at a geological site from core samples, seismic, and other limitedly available data. In addition, both the DEC and TI have configurator programs which apply procedures to information about diverse components in a system to see whether they will fit together appropriately.

Policy systems apply rules and guidelines to particular circumstances in order to determine appropriate actions. HIT is such a system. One of the first policy systems was Newell, Shaw, and Simon's chess-playing program (Newell et al., 1963). It followed metagoals such as "keep the king safe," "maintain control of the center squares," and "maintain a balance of material," whereas HIT is guided by principles such as "deescalate the intensity and emotionality of the situation" and "improve the informational and power position of the police." These metagoals find expression in the conclusion segments of each production rule and subsequently result in advice to follow which is consistent with the metarules and their underlying values. Several other recent systems are policy based: Purdue's Grain Market Analyzer, which finds the most profitable means of marketing grain; TI's Health Care Billing Advisor, which determines the hospital's billing process on the basis of the specified DRC, the severity of the care, and the extent of insurance coverage; and Computer Law Systems, Inc.'s ERISA Prohibited Transaction Exemption Analyst, which traverses ERISA law to determine legal retirement programs.

All of these application areas—diagnostic, procedural and classificatory, and policy—have the potential of frequent application and thus usually satisfy the first criterion.

Existence of Expertise

There are many complex problems for which there are no known experts and no identifiable body of expertise. These unique problems require creative one-of-a-kind responses. To develop an expert system, it is necessary to clearly identify and, if possible, circumscribe a relevant body of knowledge. The notion of expertise, practically as well as etymologically, derives from experience and from repeated inquiry, attempting, and putting to the test. This knowledge generally takes the form of what Bertrand Russell called "knowledge by acquaintance"—experiential or heuristic knowledge obtained by direct cognitive relationship and involvement with the target matter (Russell, 1917)—or "personal knowledge"—Michael Polanyi's seeming oxymoron which traces the evolution from tacit knowledge one holds private to explicit knowledge one can share publicly (Polanyi, 1958). Since most of the relevant knowledge is not stated explicitly, the persons in whose minds the expertise resides must be identified.

For our HIT application we used trained negotiations officers and SWAT team officers as the primary base of expertise. Many police departments and law enforcement agencies have identified personnel who have experience in this area (although, to anticipate a subsequent criterion, this expertise is scarce). So there is a traceable path leading from the problem to the relevant expertise. Somewhere behind every expert system lies at least one expert—a medical doctor, an equipment troubleshooter, a chemist, a geologist, a computer configurator, an accountant, or a lawyer. If you cannot identify one or more such people, developing an expert system is problematic.

Sometimes expertise lies far deeper than you think at first. Researchers at Texas Instruments learned this when they built an expert system for the cockpit of the F-16. Among the many considerations that led them to choose the F-16 project as a prototype was the fact that a detailed operating manual was available. It was assumed that the primary job would be to translate that knowledge into LISP code for execution on the Explorer. That assumption was debunked quickly. The manual tells the pilot what steps to take when a one-dimensional situation occurs, such as the loss of compressionization in the cockpit. These well-defined procedures the ES designers gobbled up quickly and translated effortlessly into rules and frames. The designers became stalemated, however, when they were forced to deal with many events at once—that is, to deal with the gist of reality. Going to real experts, in this case experienced test pilots, was the only way around this barrier. The knowledge contained in the F-16 manual was necessary, but it was insufficient by some measure to develop the system. The requisite knowledge could only be obtained by identifying an expert and reaching for his or her personal knowledge. This leads us to our next criterion.

Articulatability

The relevant knowledge must in principle be articulatable. There must be a clear chain of communication from the recesses of the expert's mind to recording mechanisms of the knowledge engineer and then to the embodiment of the knowledge in hardware and software. Considerable progress has been made during the last few years on the second stage of this sequence—from knowledge engineer to computer. Shells such as Knowledge Craft, KEE, ART, Personal Consultant, Personal Consultant Plus, and Expert System Toolkit greatly facilitate the progress. They are awkward but manageable. The first phase of the sequence, however, is sorely lacking in the development of tools and methods. Often called "knowledge elicitation," this phase continues to be largely a black art. This I believe is due to the mysteries that philosophers like Russell and Polanyi sought to understand.

Key to the success of any expert systems project is the knowledge engineer's dedication to the process of inquiry which will unravel enough of the embedded mystery to be useful and the design of an inquiry system to guide this effort. Churchman (1971) recognized this as one of the initiators of DENDRAL. One had to attack this opportunity as a problem in epistemology and with the spirit and discipline of the great inquirers. Joshua Lederberg, Edward Feigenbaum, Bruce Buchanan, and a host of chemists in Lederberg's research program at Stanford became imbued with this spirit. Early runs of DENDRAL were able to identify correctly only a very small percentage of compounds. They took that as a "win" and pressed on. There was a flurry of excitement as new knowledge was discovered, articulated, and a few more percentage points gained. In the case of DENDRAL this effort was so successful that today DENDRAL is no longer an expert system. Through research motivated in large part by DENDRAL itself, the underlying science is now known with certainty. It is now public and analytic knowledge and does not rely on the hunches and heuristics of experts. The full process of articulation has been completed.

If Michael Polanyi is correct, however, there will always be some core of inarticulable, personal knowledge underlying most expert systems. The judgment that one needs to make in choosing an expert systems project is how far it is possible to go in making explicit the expert's knowledge.

In developing the HIT system, we discovered an exceptional expert in Captain Ronald Zuniza, Commander of the Hostage Negotiation Unit of the Tucson Police Department. Zuniza viewed the prototype created from secondary sources, lauded it, and then he tore it apart. It was a good idea, he thought, but incomplete and institutionally ill fitted. So Vedder and Zuniza spent many long hours discussing, debating, and reformulating principles, ideas, and procedures. As they did, Zuniza's tacit knowledge became more explicit. Ultimately, it took on a form that could be translated into a production rule–based expert system.

To aid the knowledge-eliciting process such as Zuniza and Vedder engaged in, Alex Tsakiris and I are exploring the use of argumentation analysis of the type that Ian Mitroff and I applied to strategic planning (Mason and Mitroff, 1981). Tsakiris's prototype package ARK-4 shows promise in preliminary application at organizations such as Texas Instruments.

Scarcity of Expertise

If there are numerous occurrences, articulatable knowledge, and identifiable experts but the expertise is commonplace, an expert systems application is unlikely to be economical. The best applications generally are those in which the expertise requires extensive experience, education, and training and perhaps some special innate talents. As a consequence, the expertise is usually expensive to acquire and maintain. Stated differently, the person or persons possessing the expertise must have a high human resource value to the organization.

Using this criterion, possible application areas can be uncovered by evaluating individuals in an organization on the basis of what Eric Flamholtz calls "expected realizable value." This value consists of two components:

1. the amount the organization could potentially realize for his or her services if the person maintains an organization membership during the period of his or her productive service life, and
2. the amount actually expected to be derived, taking into account the person's likelihood of turnover.

These two factors speak to the overall economic value of the individual to the organization. Anyone who potentially will make substantial contributions to the organization, but who also has a high probability of turnover probably possesses expertise that is a candidate for an expert system.

This approach was used implicitly in one of the most heavily publicized recent expert systems, Campbell Soup's bacteria control system. Executives at the company's Camden, N.J., plant became concerned when they realized that Aldo Cimino was approaching retirement age. Aldo was 63 years old, had worked for Campbell's for 43 years, and during the last two decades had become the company's key expert on "cookers"—those five-story vats in which soup is brewed. There are more than 90 of these cookers in Campbell's factories around the world. Cimino had become the diagnostic expert for these cookers and was in charge of troubleshooting, maintenance, and repair and also participated in the design and operation of the equipment. The cookers contain hydrostatic and rotary sterilizers that are used to kill botulism bacteria in canned soup—Campbell's major product. So Cimino's expertise was central to the performance of the company as a whole, and for that reason he was constantly traveling to meet or talking on the telephone with people at the 90 sites to help them solve problems. His loss posed a real threat to Campbell's. In response, the company set about to construct an expert system comprised of some 150 rules, using Texas Instrument's Personal Consultant system. The resultant system makes about 90% of Cimino's scarce knowledge about cookers available to 90 sites.

HIT illustrates other dimensions of scarcity. Most large departments and agencies can afford only a very few specially trained negotiation and crises officers. The vast majority of law enforcement organizations cannot afford the training and other costs involved. Yet hostage-taking situations are unpredictable. Although they tend to cluster in certain types of establishments such as banks and convenience stores, they can occur any place at any time. Thus the distribution of expertise is sparse compared to the spatial and temporal probabilities of occurrence of events requiring expert response.

Many of the cost-effective expert systems create value by capturing limited expensive expertise, which is required in dispersed locations at different times, making it more broadly available.

Negligence

Bruce Buchann once commented that expert systems will have "made it" when, say, a physician will be held liable for negligence for failing to use an expert system such as MYCIN when treating a patient, that is, when an expert system has become one of the state-of-the-art tools that a prudent practitioner is expected to use in the professional conduct of his or her business.

Although the previous criteria are related to the feasibility and positive benefits from employing expert systems, this one relates to the costs incurred from failing to apply the expertise. If failure to apply the right expertise in a situation can result in significant losses, the area is a candidate for expert systems. Further, if those responsible can be held accountable for negligence of duty if the appropriate expertise is not employed, even a stronger case can be made. In posing this criterion, I have not dealt with the issues of validation and acceptance. These represent crucial problems to be solved before a system can be successfully implemented. Nevertheless, if this criterion is satisfied, the area represents a good potential application for expert systems.

In the case of hostage-taking incidents, numerous volatile cases occur each year. For most of these incidents the scarce expertise of special negotiation officers is not available; the consequence is death or injury to one or more persons. Inability to apply the expertise in these cases results in significant social losses.

Failure to apply Aldo Cimino's expertise could result in the loss of an entire batch of soup, or worse, the widespread infestation of botulism bacteria in cans of soup. At stake with the F-16 support system is the life of the pilot and the fate of the plane. Most existing expert systems have not yet reached this level of dependency. This is, however, the long-term goal of those who create these systems.

ETHICS

A recent issue of *Personal Computing* featured several articles and sidebars on expert systems. Among these was a short piece on the aforementioned Aldo Cimino at Campbell's. The headline read "An Expert Whose Brain Was Drained" and the body contained equally chilling metaphors about "canning" his accumulated knowledge into a personal computer expert system (Abatemarco, 1985). It appears that Cimino was an informed and eager participant in the development of the system that captured in a few weeks the knowledge it had taken him a lifetime to acquire. He probably entered into a just and fair bargain. So there is probably no major ethical issue involved in his case.

The expressed idea of draining one's brain does not, however, rest comfortably with my moral precepts. It carries the sense of a deep violation of something private and personal. It is a threat to one's dignity, especially in the twentieth century and in an information age in which one's mind has become such a central part of one's being. The prospect of draining one's being therefore takes on strong moral overtones.

In developing an expert system we proceed by extracting knowledge from experts, workers, and other knowledgeable people. Then the expertise is implanted into computer software, where it becomes capital in the economic sense. This process of "disemminding" knowledge from an individual—akin to the disembodiment of energy from workers that took place during the industrial revolution—and the subsequent process of "emminding" it into computers have the sociopolitical effect of transferring control of one's intellectual property to those who own the hardware and software. This transfer of property raises many moral and ethical questions, questions that should be at the forefront of any assessment of expert systems technology. Is the transfer of property warranted? Just? Did it take place under conditions of truly informed consent? Who owns the knowledge? To what extent is it private or public? How is the contributor of the knowledge to be compensated? There are among the questions of human values that must be answered as we create more intelligent information systems.

CONCLUSION

This leads me to a summary statement of the general criteria by which one should evaluate a potential expert systems application. The knowledge involved should exist in the mind of one or more identifiable experts. It should be articulatable and extractable. It should be relatively scarce, yet applicable to many different situations. And the failure to apply the knowledge in circumstances that require it should result in significant losses. Most important, the transfer of knowledge from the expert to the owner and users should be just and morally defensible. As economic and social forces require more executives to consider expert systems seriously as a matter of competitive advantage or survival, they are well advised to keep these evaluation criteria in mind.

REFERENCES

ABATEMARCO, FRED (1985). An expert whose brain was drained, *Personal Computing* (November):98

CHURCHMAN, C. WEST (1971). *Design of Inquiring Systems,* Basic Books, New York.

MASON, RICHARD O., and IAN I. MITROFF (1981). *Challenging Strategic Planning Assumptions,* Wiley, New York.

NEWELL, ALLAN, J. C. SHAW, and H. SIMON (1963). Chess-playing programs and problems of complexity, in Edward A. Feigenbaum and Julian Feldman (eds.), *Computers and Thought,* McGraw-Hill, New York, pp. 39–70.

POLANYI, MICHAEL (1958). *Personal Knowledge,* University of Chicago Press, Chicago.

RUSSELL, BERTRAND (1917). *Mysticism and Logic,* Allen & Unwin, London.

STRENTZ, THOMAS (1983). The inadequate personality as a hostage taker, *Journal of Police Science and Administration* (September):363–368.

VEDDER, RICHARD G., and Richard O. Mason (1986). An expert system application for decision support in law enforcement, *Working Paper ELCSB-MIS,* Southern Methodist University, Dallas, Tex.

INDEX

D

E

F

G

I

L

M

O

P

R

S

T

U

V

W

X